Natural Environment Research Council

Institute of Terrestrial Ecology

A Field Key for Classifying British Woodland Vegetation Part 2

R G H Bunce

London: Her Majesty's Stationery Office

ISBN 011 701417 6

Cover design C B Benefield

The Institute of Terrestrial Ecology (ITE) was established in 1973, for the former Nature Conservancy's research stations and staff, joined later by the Institute of Tree Biology and the Culture Centre of Algae and Protozoa. ITE contributes to, and draws upon, the collective knowledge of the 14 sister institutes which make up the **Natural Environment Research Council**, spanning all the environmental sciences.

The Institute studies the factors determining the structure, composition and processes of land and freshwater systems, and of individual plant and animal species. It is developing a sounder scientific basis for predicting and modelling environmental trends arising from natural or man-made change. The results of this research are available to those responsible for the protection, management and wise use of our natural resources.

One quarter of ITE's work is research commissioned by customers, such as the Department of Environment, the Commission of the European Communities, the Nature Conservancy Council and the Overseas Development Administration. The remainder is fundamental research supported by NERC.

ITE's expertise is widely used by international organizations in overseas projects and programmes of research.

Dr R G H Bunce
Institute of Terrestrial Ecology
Merlewood Research Station
GRANGE-OVER-SANDS
Cumbria
LA11 6JU

05395 (Grange-over-Sands) 32264

CONTENTS

INTRODUCTION

The original project on which this Field Key is based was set up in order to produce a woodland classification for what was, at that time, the Nature Conservancy. The conservation branch, now the Nature Conservancy Council, required a classification for the identification of key sites, and the research branch, now the Institute of Terrestrial Ecology, needed to fit research areas into the national context.

In the early discussion, the main emphasis was laid upon the classification of whole sites. At that time, a survey of woodland sites was in progress for the *Nature conservation review* (Ratcliffe 1977), wherein lists of a limited range of species were recorded. Accordingly, these data were used in a preliminary classification to test the approach to the analysis of site complexes. In the process of these studies, it was established that the majority of users were more familiar with the vegetation within plots, rather than being able to accept the vegetation complexes contained within entire woods (or sites as they have been termed in the present context in order to emphasize the 2 levels involved). The decision was therefore made to formally separate these 2 generally used levels into separate classifications that were complementary to each other. The degree of abstraction is arbitrary, and no evidence has yet been found to show any actual discontinuity.

Although the primary objective of the study was to classify whole woods, the first stage of the classification has been published in Part 1 in order to provide the background to the site classification. The plot classification in Part 1 provided a dichotomous key to identify 32 types of woodland vegetation at a unit scale of 200 m^2. The present booklet is concerned with a treatment of plots as samples representing whole sites, to produce a classification which overtly uses the heterogeneity present within the woods as a basis for their classification. In general, British woodlands are fragmented and, therefore, are usually treated as units for study and for evaluation purposes. On the Continent, these units would be comparable to compartments. Although there are many informal site classifications, eg valley woodlands of the west, there is no formal system comparable to that described in the present booklet. By contrast, there are several classifications available at the level of vegetation types, eg the stand types of Peterken (1981) and the phytosociological series described by Klötzli (1970).

It has frequently been argued in such texts that uniform areas of vegetation need to be identified in order to simplify the classification process. However, the treatment of whole woodland sites together inevitably requires sites to be accepted as complexes of vegetation types, thus rendering traditional procedures inapplicable.

The main objective is concerned with the concept of representativeness which is central to the *Nature conservation review* (Ratcliffe 1977), and forms the basis of site selection as notified areas. Inevitably, there is confusion between scientific criteria for selection and subjective evaluation, as described by Margules and Usher (1984). Whilst the analogy has been drawn between the identification of important sites as part of a cultural heritage (Tittensor 1981), it is generally recognized that scientific criteria should take precedence, as emphasized by Bunce (1981), and that classification should precede evaluation. This point of view is admirably summarized by Austin and Margules (1984), who stated that the assessment of representativeness requires:
- a hierarchical system of ecological units;
- a definition of the relevant properties of such units;
- a method of allocating potential reserves to such units; and
- a means of evaluating the representativeness of such units.

The classification described below meets all of these criteria. It is the first document to enable whole woodland sites to be classified, and thus provides a basis for evaluation in that sites can be compared despite the heterogeneity of their vegetation.

The preliminary analysis, described in Part 1, was based on species lists from entire sites, and served as a basis for the second stage of the project which involved detailed surveys from 103 woodlands. Part 1 was designed to explore the variation within the constituent vegetation, and it resulted in the Plot Key. Several other factors also favour the entire site approach – for example, the scale is suitable for larger animal studies, because they are not confined to individual vegetation units but move across several types. The present Key identifies the affinities of entire sites in the national context, and may be used to provide the background for studies other than those directly concerned with vegetation, eg animal behaviour. The series may also be used as basis for site selection where a range of variation is required, eg for comparison of *Betula* (birch) populations. The studies may then be set in the national context. The species constituting the vegetation are used as a means of obtaining an integrated picture of site character rather than as an end in themselves, as in the phytosociological tradition developed by Braun-Blanquet. Thus, as explained in Part 1, the entire species composition is used, rather than the tree species of the canopy alone, because there is then more information available. The canopy has invariably been modified by management, and is an unreliable indicator without other peripheral information. The site classification, therefore, represents the overall biological affinities of the sites as reflected by the integration of species complexes.

Visual criteria are minimized in the classification because they are largely based on the perception of the individual observer and cannot therefore be used as a reliable means of assessment. It must be expected that, in some cases, widely different sites in visual terms may be assigned to a similar woodland type because the convergence of site conditions gives rise to similar species combinations. For example, a sloping site on the west may be comparable to a shallow site in the east.

It must therefore be recognized that the site classification represents a break from traditional systems of vegetation classification, in the size of scale employed. Its use represents a challenge in accepting a higher degree of abstraction than is required in traditional systems and in involving the acceptance of heterogeneity within the sites.

Inevitably, because of the intensity of sampling required in each woodland, there are relatively few (103) sites on which to base the classification. Some of the types are therefore defined by few sites, but, in the tests shown below, it is demonstrated that the classification is stable and that individual sites generally fit into the site type definitions. However, the classes may be too generalized for local studies – in which case, as in the native woods of *Pinus sylvestris* (Scots pine), further survey may be required to define local types. Surveys of woods in north Wales and southern Scotland have demonstrated the way in which such supplementary surveys need to be carried out. The Key assigns woodlands to 15 types representative of the range of variation in Britain. In contrast to phytosociological classification, these types should not be considered final, but rather a step towards a progressively refined system.

Within the context of the survey, the following definition of a woodland site was applied:

an area of woodland over 5 ha, in which exotic species have a representation below 25%, defined by boundaries obviously present on a map.

Within Britain, the above definition applies to most areas of semi-natural woodland, and it is only occasionally difficult to separate continuous blocks of forest. Where such blocks need to be divided, decisions should be made on cartographic criteria, rather than on observed differences in vegetation. Fragments of woodland separated from a whole catena by arbitrary felling may thus possibly come into different site types – because they contain different complexes of vegetation types, and hence have different characteristics and occupy different environments. As with all such generalized classifications, unique sites occupying particular sites of unusual environmental combinations, eg sea cliffs on limestone, will not be identified. However, they will be assigned to a closely related type and can be defined subsequently by their unique features, usually in terms of habitat.

To reiterate, therefore, the classification concerns the 'representative' criterion of the *Nature conservation review* (Ratcliffe 1977). It is recognized that additional criteria for conservation, such as the age of the woodland, need to be determined in addition to the primary factor. Thus, 2 sites, say in East Anglia, may have the same basic complement of vegetation types but could be of very different ages, with different detailed species composition. Sixteen samples per site were chosen because it was found, after resampling several sites, that a high degree of reproducibility was achieved, even on variable sites. Since the main survey was completed, 8 plots have been shown to be sufficient in most cases to establish definitively the type to which the site belongs. As with the plot classification, experience has been gained throughout Britain to demonstrate that the Key works efficiently. Previously collected data can also be used, provided that the overall estimate is not biased by selective sampling – the frequency records will need to be converted to categories comparable to those used in the Key, ie a basis of 16 samples per site. The attendant risks of misclassification increase as the divergence increases from the initial methodology.

The method used to select the sites for detailed study described in Part 1, ie an analysis of species data from 2463 woods throughout Britain, was intended, as far as possible, to cover the full range of variation within British woodlands. Further sites surveyed have fitted into the range as defined, thereby suggesting that the sample was representative. However, unusual features of particular sites are more readily appreciated than patches of vegetation at the plot scale. The site classification takes the overall weight of the species composition of the site, and places it in a type of the classification accordingly. The analysis identifies common features rather than differences, and individual peculiarities are minimized.

The commoner types are certainly covered adequately, but there remains the possibility that the rarer types have not been identified. By definition, a type needs to be defined by constantly recurring characteristics, and can thus only be identified by a detailed regional study. It is not, therefore, sufficient to identify individual sites and to establish a new type on the basis of their differences. Rather, a series of sites will need to be surveyed, using a comparable procedure to the original survey, and a new analysis carried out in order to produce a new Key – as has already been carried out for the native pine woods. In this case, the analysis expands the existing Key, as opposed to creating new types outside the present range. Small sites may be included, but, if outside the range of definition of the sites originally included, it will not be surprising if they fall outside the described range of types, because the scale is arbitrary and small sites will eventually occupy one plot vegetation type. Plots selectively placed in order to investigate specific areas of vegetation should not be used to obtain frequency data for the Site Key, as they will introduce a bias to the sampling. Plots used to obtain data for input to the Site Key should be positioned so as to obtain an unbiased estimate of overall frequency, or at least an effort should be made to obtain such an estimate from extant data.

Although the experience to date suggests that the Key is robust, it is a risky procedure to estimate by eye the frequency of species in sites, as it not only involves an intimate knowledge of the site, but also an ability to balance the mosaics present in any woodland. However, provided that the site is relatively uniform, or distinct in its species composition, it is recognized that it can be ascribed to an appropriate type with a high degree or probability of success. Such short cuts in the full procedure may be necessary for practical reasons, and can provide an insight into site relationships relatively quickly. In comparison with the plot classification, where it is difficult to envisage a detailed species complement, it seems likely that a good estimate of the position of the site may be obtained, provided that the observer knows a whole site rather than a few well-known patches.

Regional studies may well be carried out in order to develop more detailed Keys for particular areas, as for the native pine woods (Bunce 1977). The approach has, therefore, a major advantage in that it is capable of expansion into a range of detailed studies. For example, 2 mixed deciduous woodlands in western Scotland, Glen Nant and Glasdrum, have many points of difference, yet came into the same type. A more detailed survey of western Scottish woods would produce a classification which would undoubtedly separate these sites – however, the base-line classification is still needed in order to place the sites into a national context. The method of classification for evaluation has been used in the Yorkshire Dales and Northumberland National Parks, and the principles are described by Pilling et al. (1979).

In the original survey of the 2463 sites, there was broad correspondence with the degree of tree cover and the number of sites surveyed in a given region. As a result, because Scotland had a comparatively small woodland area, correspondingly fewer sites were surveyed. However, analysis of the site data has shown that these sites are very distinctive – as demonstrated by the small number of woods falling within the predominantly Scottish types. In particular, type 13 is based on only 4 sites; initially, this type was split into 2 groups, but inadequate definition was provided by the limited number of sites. Hence, it was decided to join the original 2 sets together as site type 13, and accordingly reduce the total number of types to 15 rather than the initial 16. The pine wood sites already have a subsidiary Key available (Hill et al. 1975), but a full assessment of their relaticnship with other western Scottish sites requires further survey.

With 10 indicator species at each stage, the Key is likely to be sufficiently robust to cope with some misidentification, and, with frequency being used, these errors are less likely to be serious. Such errors are likely to be more important in sites that are on the border-line between types. The reliability of estimating frequencies has already been discussed. Several studies have been carried out on this subject, and are summarized below.

Tyndrum	Two surveys by independent observers with different random co-ordinates for the plot: both sets were classified as type 13.
Glasdrum	Four sets of 16 randomly located plots: all sets were classified as type 15.
Glen Nant	Four sets of 16 randomly located plots: all sets were classified as type 15.
Craigellachie	Two sets of 16 randomly located plots: both sets were classified as type 13.
Wood of Cree	Five sets of 16 randomly located plots: 4 sets were classified as type 11 and one as type 14.

The data from Glasdrum, Glen Nant and Wood of Cree were provided by A D Horrill and J M Sykes.

The first 4 sites produced results that are stable in the classification – knowledge of the sites suggests that they are central members of the types to which they belong. Both Glasdrum and Glen Nant would generally be regarded as being heterogeneous in terms of both species and habitats, but the heterogeneity is distributed relatively evenly throughout the sites. In contrast, in the Wood of Cree, the variation is localized, and caused by wet flushes in particular areas. The degree of representation of these flushed areas in the sample tips the balance into the different types. However, these types are closely related to one another in the classification, and the site is therefore on the border-line between them.

All the above are upland sites and would be widely regarded as being highly variable. So far, the same degree of attention has not been paid to the lowland sites, many of which are more uniform than their upland counterparts. In these cases, repeated sampling will inevitably classify the sites as belonging to the same type, regardless of the number of samples. For example, site 27 had 16 plots belonging to the same plot type and is very homogeneous. However many replicate samples were put into this site, it would always end in the same type.

The majority of sites, therefore, contain reproducibly consistent assemblages of species, and even those that represent a more varied picture produce comparable results. If the site is genuinely border-line, then the classificatory position to which it is assigned will depend upon which aspect of its variation is emphasized in the particular example. As with the Plot Key, however, a given species list will always end up in the same type.

ASSESSMENT OF RELATIONSHIPS BETWEEN PLOT AND SITE CLASSIFICATIONS
As many readers will already be familiar with Part 1, and because the objective of the present publication is to integrate the plots with the descriptions at the site level, it is useful to provide a general discussion of the structure of the analysis. In many respects, the divisions made in the site classification are analogous to those made at the plot level, because the basic data are the same but recombined into complexes at a higher level of abstraction. Comparable major environmental trends underlie the divisions, as they determine the vegetation at whatever level is considered. The first division is thus virtually parallel but the second division is related in a more complex way, with site types 5–8 being comparable with plot types 1–8 and site types 1–4 with plot types 9–16. The latter parts of the analysis are in a similar order. The site types contain complexes of plot types, as expressed by the mixture provided in the descriptions. Direct use of the plot types alone with the classification would have simplified the separation of types artificially, and it was considered closer to the real situation to use the entire species complement and to accept the more complex classification that results.

Site types 1–4 contain, in general, an average number of plot types, in which 9, 10 and 12 are particularly prominent. The vegetation is quite heterogeneous, with an average number of species present. In contrast, site types 5–8 are not hetero-geneous and have a relatively narrow range of species, with relatively few plot types, 5 and 7 being prominent. Site types 9–12 are much more variable, with plot types 25, 21 and 23 commonly present. Site types 13–16 are the most variable and contain a wide range of plot types, in particular 26, 27 and 28, as well as a wide range of species.

The constant species show a similar pattern, *Rubus* (bramble) dominating the earlier site types, but giving way to a wide range of species in types 5–8. Types 9–12 have few constant species, whereas the series 13–16 has a wide range of species consistently present. *Rubus* also forms the main ground cover in the earlier types, becoming codominant with *Mercurialis perennis* (dog's mercury) in types 5–8. In the second half of the series, the dominants are consistent because relatively few species contribute overall.

The selective species reflect closely the individual features of the site types, and a knowledge of their ecology can provide useful information concerning the relationships. Thus, the earlier types have species such as *Fagus* (beech) and *Lonicera periclymenum* (honeysuckle) as selectives, whereas types 5–8 have species such as *Acer campestre* (field maple) and *Galeobdolon luteum* (yellow archangel). In sharp contrast, types 9–12 are characterized by species such as *Digitalis purpurea* (foxglove) and *Teucrium scorodonia* (wood sage). The final section has species such as *Trientalis europaea* (chickweed wintergreen) and *Galium saxatile* (heath bedstraw), reflecting the upland and acidic nature of the sites.

The tree species follow similar patterns, again emphasizing the close correspondence between the various ecological features of the vegetation when treated on a broad scale. Thus, site types 1–4 have *Quercus* (oak) mainly as a constant tree species, whereas types 5–8 have *Fraxinus* (ash) associated with oak and other trees such as field maple. In this half of the series, the species of oak is probably mainly *Q. robur* (pedunculate oak) whereas the second half is associated mainly with *Q. petraea* (sessile oak), and with birch.

Comparable features are therefore used to characterize the site types, as for the plot types. Because sites are variable on a different scale than plots, a great deal of variation is included within the individual site types.

The relationship between environment and the vegetation has already been discussed, with the principal trends being identical at either level, plot or site. Accordingly, a brief introduction to the relationships between the major groups of types is given below.

As with the plot classification, the first major division is between the lowland woods of the south and east growing under low rainfall conditions, with generally shallow slopes and relatively high pH (with the exception of types 1 and 2), and the woods of the north and west (with the exception of type 9) growing under high rainfall conditions, with steep slopes and low pH. Types 1–4 are more acidic than 5–8, and also tend to be on steeper slopes in a wider range of geographical locations, because types 5–8 are mainly in East Anglia and the south-east. The extreme types 13–16 are characteristic of the Highlands and the north of England. Types 9–12 are more widespread, but are on steep slopes and occur under a wide range of landforms.

As with the plot types, the next stage is to consult the detailed definitions of each of the types. In the case of the sites, this definition is probably even more advisable as it is easier to have an overall impression of a site than a plot. It is important to bear in mind that the classification places the sites according to their overall affinities and the balance of species present. On first appearance, a site may be wrongly placed, but a closer examination of the species composition and plot type characteristics will probably reveal important similarities, as it is often difficult to give a complete definition of all the site environmental features.

WORKED EXAMPLE

The frequency of the following species was obtained from records made in 16 randomized plots in a wood near Alvie in the central Spey valley in Highland Region. In contrast to Part 1, the frequency of the species is important as the values are used in the Key. Species with no English names attached are bryophytes and follow flowering plants at all stages in the text.

Achillea millefolium (yarrow)	1
Ajuga reptans (bugle)	1
Agrostis canina (brown bent-grass)	13
Agrostis tenuis (common bent-grass)	12
Anemone nemorosa (wood anemone)	1
Anthoxanthum odoratum (sweet vernal-grass)	8
Arrhenatherum elatius (oat-grass)	2
Betula spp. (birch)	4
Blechnum spicant (hard-fern)	4
Brachypodium sylvaticum (slender false-brome)	4
Calluna vulgaris (ling)	13
Campanula rotundifolia (harebell)	2
Carex binervis (ribbed sedge)	1
Cerastium holosteoides (common mouse-ear chickweed)	3
Conopodium majus (pignut)	1
Dactylis glomerata (cock's-foot)	2
Deschampsia flexuosa (wavy hair-grass)	16
Digitalis purpurea (foxglove)	4
Dryopteris filix-mas (male-fern)	3
Empetrum nigrum (crowberry)	2
Erica cinerea (bell-heather)	4
Erica tetralix (cross-leaved heath)	1
Festuca ovina (Sheep's fescue)	12
Galium saxatile (heath bedstraw)	12
Galium verum (lady's bedstraw)	1
Hieraceum pilosella (mouse-ear hawkweed)	1
Holcus lanatus (Yorkshire fog)	7
Holcus mollis (creeping soft-grass)	2
Juniperus communis (juniper)	1
Lathyrus montanus (bitter vetch)	1
Lolium perenne (rye-grass)	1
Lotus corniculatus (birdsfoot-trefoil)	2
Luzula multiflora (many-headed woodrush)	14
Luzula pilosa (hairy woodrush)	1
Luzula sylvatica (greater woodrush)	1
Mercurialis perennis (dog's mercury)	1
Molinia caerulea (purple moor-grass)	3
Nardus stricta (mat-grass)	1
Oxalis acetosella (wood-sorrel)	7
Pinus sylvestris (Scots pine)	1
Poa pratensis (meadow-grass)	2
Polygala serpyllifolia (common milkwort)	1
Potentilla erecta (common tormentil)	11
Primula vulgaris (primrose)	2
Prunella vulgaris (self-heal)	1
Pteridium aquilinum (bracken)	5
Pyrola media (intermediate wintergreen)	1
Rubus idaeus (raspberry)	2

Rumex acetosa (sorrel)	2
Sarothamnus scoparius (broom)	1
Sieglingia decumbens (heath grass)	4
Sorbus aucuparia (rowan)	4
Stellaria alsine (bog stitchwort)	1
Succisa pratensis (devil's-bit scabious)	2
Teucrium scorodonia (wood sage)	4
Thelypteris dryopteris (oak fern)	2
Trifolium repens (white clover)	1
Urtica dioica (stinging nettle)	1
Vaccinium myrtillus (bilberry)	11
Vaccinium vitis-idaea (cowberry)	8
Veronica chamaedrys (germander speedwell)	4
Veronica montana (wood speedwell)	2
Viola riviniana (common violet)	11
Dicranum fuscescens	2
Dicranum majus	1
Dicranum scoparium	3
Hylocomium splendens	15
Hypnum cupressiforme	4
Lophocolea bidentata	2
Plagiochila asplenioides	7
Pleurozium schreberi	10
Polytrichum spp.	2
Pseudoscleropodium purum	8
Plagiothecium undulatum	6
Rhytidiadelphus loreus	5
Rhytidiadelphus squarrosus	3
Rhytidiadelphus triquetrus	10
Sphagnum spp.	2
Thuidium tamariscinum	2
Trichocolea tomentella	1

Having prepared the species frequency, it is now appropriate to consider STEP ONE of the Key (Page 21). For convenience, species in the left-hand column are printed in upper case, while those in the right-hand column are printed in lower case. The following frequency classes need to be considered:

Class		Code
1– 4	=	1+
5– 8	=	5+
9–12	=	9+
13–16	=	13+

A species with a frequency of 7 will therefore score in the 5–8 class, as represented by 5+ in the Key, whereas a frequency of 11 will be represented by 9+. If a species with these codes is present in the left-hand column, it scores – 1, ie it counts negatively; if a species from the right-hand column is present, it scores +1 ie it counts positively. A species which scores 13+ is also considered as scoring 1+, 5+ and 9+, as the analysis was originally described by Hill *et al.* (1975). Before keying out the frequencies from Alvie, it is desirable to consider some hypothetical examples to indicate how the balance between negative and positive species frequencies is obtained using STEP TWO as an example. For a species to count in the Key, its score must be at least that specified. If 5+ is given, then scores of 5+ or more would count; 1+ would not.

Negative		*Positive*	
BLECHNUM SPICANT	1+	*Mercurialis perennis*	13+
(HARD-FERN)		(dog's mercury)	
DIGITALIS PURPUREA	1+	*Urtica dioica*	9+
(FOXGLOVE)		(stinging nettle)	
OXALIS ACETOSELLA	5+		
(WOOD-SORREL)		*Thamnium alopecurum*	1+
PTERIDIUM AQUILINUM	5+		
(BRACKEN)			
SORBUS AUCUPARIA	1+		
(ROWAN)			
PELLIA SPP.	1+		
POLYTRICHUM SPP.	1+		

Score −1 or less (ie −1, −2, −3, . . .), go to STEP THREE
Score 0 or more (ie 0, +1, +2, . . .), go to STEP SIX

Five hypothetical combinations of indicator species frequencies are shown below, with their appropriate scores.

a. BLECHNUM SPICANT
    (HARD-FERN)   4 ie 1+ −1
    OXALIS ACETOSELLA
    (WOOD-SORREL)   13 ie 13+ −1   } = −3   (ie Go to STEP THREE)
    SORBUS AUCUPARIA   (less than the threshold of −)
    (ROWAN)   4 ie 1+ −1

b. BLECHNUM SPICANT
    (HARD-FERN)   6 ie 5+ −1
    OXALIS ACETOSELLA
    (WOOD SORREL)   13 ie 13+ −1
    SORBUS AUCUPARIA   } = −3+1 = −2   (ie Go to
    (ROWAN)   4 ie 1+ −1   STEP THREE)
    URTICA DIOICA   (equal to the threshold of −1)
    (STINGING NETTLE)   10 ie 9+ +1

                                            (If the score is equal to the threshold,
                                            then also go to STEP THREE)

c. *OXALIS ACETOSELLA*   13 ie 13+ −1 ⎫
   (WOOD-SORREL)   ⎬ = +1−1 = 0   (ie Go to STEP SIX)
   URTICA DIOICA   10 ie 9+ +1 ⎭ (more than the threshold of −1)
   (STINGING NETTLE)

d. *OXALIS ACETOSELLA*   13 ie 13+ −1 ⎫
   (WOOD-SORREL)   ⎪
   URTICA DIOICA   10 ie 9+ +1 ⎬ = −1+2 = +1   (ie Go to
   (STINGING NETTLE)   ⎪   STEP SIX)
   MERCURIALIS PERENNIS   13 ie 13+ +1 ⎭ (more than the threshold of −1)
   (DOG'S MERCURY)

e. *URTICA DIOICA*   10 ie 9+ +1 ⎫
   (STINGING NETTLE)   ⎪
   MERCURIALIS PERENNIS   13 ie 13+ +1 ⎬ = +3   (ie Go to STEP SIX)
   (DOG'S MERCURY)   ⎪ (more than the threshold of −1)
   THAMNIUM ALOPECURUM   3 ie 1+ +1 ⎭

Now consider the frequency data for the wood at Alvie. On inspecting the species list for the indicator species in STEP ONE, the following should be identified:

STEP ONE
(Page 21)

INDICATOR SPECIES

Negative	*Positive*	
	Agrostis canina	13 ie >1+ +1
	(brown bent-grass)	
	Agrostis tenuis	12 ie >5+ +1
	(common bent-grass)	
	Anthoxanthum odoratum	8 ie >1+ +1
	(sweet vernal-grass)	
	Deschampsia flexuosa	16 ie >5+ +1
	(wavy hair-grass)	
	Galium saxatile	8 ie >1+ +1
	(heath bedstraw)	
	Vaccinium myrtillus	11 ie >1+ +1
	(bilberry)	
	Dicranum scoparium	3 ie 1+ +1
	Plagiothecium undulatum	6 ie >1+ +1

Total score +9, therefore proceed to STEP NINE

13

By referring to Page 29 and the species frequency, the following indicators were identified for this step:

STEP NINE
(Page 29)

INDICATOR SPECIES

Negative   Positive

Agrostis canina	13 ie	>5+ +1
(brown bent-grass)		
Calluna vulgaris	13 ie	>5+ +1
(ling)		
Galium saxatile	12 ie	9+ +1
(heath bedstraw)		
Luzula multiflora	15 ie	>5+ +1
(many-headed woodrush)		
Nardus stricta	1 ie	1+ +1
(mat-grass)		
Polygala serpyllifolia	1 ie	1+ +1
(common milkwort)		
Potentilla erecta	11 ie	9+ +1
(common tormentil)		
Hylocomium splendens	15 ie	>5+ +1
Pseudoscleropodium purum	8 ie	5+ +1

Total score +9, therefore proceed to STEP THIRTEEN

By referring to Page 33 and the species frequency, the following indicators were identified for this step:

STEP THIRTEEN
(Page 33)

INDICATOR SPECIES

Negative   Positive

CALLUNA VULGARIS   13 ie 13+ −1
(LING)

Total score −1, therefore the wood at Alvie is representative of SITE TYPE 13, upland woods on freely drained sites.

On inspection of the description, it will be seen that the species list conforms to the array of 'constant' and 'selective' species given. It should also be noted that many of the species given in the complete list for the site were not used in identifying the site type. The smaller number of sites included in the analysis, compared with the relatively large number of plots used in part 1, does mean that the types are not based on such reliable average values, and hence there is more chance of an individual site being somewhat different from the description. It is also rather easier to attach preconceived intuitive labels in comparison with more anonymous patches of vegetation. On the other hand, experience with the Site Key suggests that it defines affinities successfully, with the use of frequency being important as it reduces the possibility of chance in the indicator species. It should be noted that a species may have a negative score at one step and a positive score at another. For example, *Sorbus aucuparia* (rowan) at a frequency of over 5+ is positive at STEP ONE, but negative with frequency of 1+ at STEP TWO.

Certain sites occupying unique situations, eg Cadgewith elm wood on a sea cliff in Cornwall or Wistman's wood high on Dartmoor, cannot be expected to fit exactly with the generalized descriptions of the site types. However, unless it is deemed preferable to label these types according to the unique character of the site they occupy, it is not appropriate to create a new site type on the basis of individual sites. Rather, a series of sites will need to be surveyed, as for the native pine woods, to create a subclassification of a particular series of sites. Whilst it is a risky procedure to estimate by eye the frequencies of species in sites, it is recognized that, where a site is not border-line, then experience of likely frequencies can ascribe a site to its appropriate type with a reasonable degree of certainty, and such a procedure may be necessary for practical reasons. In all but one case of repeated sampling of the same site, the series of 16 samples were classified as the same site types.

TYPE DESCRIPTIONS

In the type descriptions, plant species are arranged in different groupings in parallel with résumés of physical habitat attributes. These groupings have been selected to give a balanced picture of the type, rather than overwhelming the user with too many data from which further synthesis would be required. The groupings are defined below.

Each type has been given an appellation, for example:

SITE TYPE 1
PTERIDIUM AQUILINUM/QUERCUS-FAGUS TYPE (BRACKEN/OAK-BEECH)

SITE TYPE 2
HEDERA HELIX/QUERCUS-FRAXINUS TYPE (IVY/OAK-ASH)

Whereas trees were not included in the names in Part 1, it was decided that the fewer numbers of types and the scale of the site justified their use here. Furthermore, the names could be distinguished readily from those used for the plot types. However, as in the case of the example of the wood from Alvie, the major tree species from the site under study will not always appear in the name – although, in that case, it was abundantly present as saplings and with a high constancy. The names are in 2 parts: first is the ground flora selective species (see below) that occurred in over 75% of the replicate plots of the particular type; the second is composed of the 2 tree species with the highest basal areas (see below).

In the plot classification, the heterogeneity within the plot types was defined in terms of species groups determined by classifying data from the one m^2 plot size. However, site heterogeneity should be examined in terms of the combinations and

15

frequencies of plot types already described in the plot classification, and further descriptions of these types are therefore provided. The text following the Key is thus in the order given below.

1. The general description of the site types is derived primarily from the detailed data summaries, with the same objective as the brief descriptions of the plot types provided in Part 1. Because of the difference between plot and site scales, it has been made more of a narrative, although the terms are derived from numerical data. The plot types were selected as being consistently represented through the sites, and may, therefore, differ from the average occurrence given in the summaries. The additional data from surrounding land use and grazing patterns were derived from the site records described by Bunce and Shaw (1973). The relationships with Continental sites were derived from personal experience.

2. On the top of the facing page, a photograph from a site drawn at random from each of the site types is given. It must again be emphasized that individual sites may diverge widely from the particular sites shown because of convergence, ie a steep, well-drained site in the west may have comparable conditions to a level, poorly drained site in the east. It is not possible to represent adequately the variation within the types, but the photographs do give a visual impression of the general conditions prevailing.

   The map shows the predicted distribution of additional sites. In the initial stratification, woods were drawn at random from the original 103 groups (see Part 1). In order to indicate the further occurrence of woods in addition to the selected sites, the sites from each group were reclassified into the site types and added together for the 100 km squares of the National Grid for the purposes of representation. These predicted distributions give a better general picture of the likely extent of further sites than the sample of 103 sites, because use is made of the representative nature of the initial selection process.

3. Summaries of the 4 most abundant plot types present in the woods are then given. These summaries have been derived from the descriptions in Part 1, to convey the maximum information in the space available. They give a general impression of the ground vegetation, both in terms of its species composition and associated environment.

4. The final numerical summaries have the same arrangement as in Part 1, with ground vegetation information being provided first, followed by the tree and environment data.

The categories used in the site summaries are shown below.

1. **Vegetation** – 2 divisions, ground flora and woody perennials, each of these being subdivided.

1.1 **Key species of ground flora**
1.1.1 **Constant species**   These species occurred in more than 70% of plots that were present within the sites surveyed belonging to the types.

16

1.1.2 **Plot dominants** These species have an estimated cover of 5% or over within the sites surveyed belonging to the types.

1.1.3 **Selective species** To establish if a species occurred differentially, its observed frequency within a type was compared with its mean frequency over the whole series of types. The chi-square test was used to assess the departure from randomness, and the 6 species with probabilities of over 99.9% are listed.

1.1.4 **Plot types** In Part 1, the ground flora species were classified into groups according to their associations with one another. However, at the site level, the various combinations of the plot types described in Part 1 are more appropriate for assessing the composition of the flora. The average number of species groups present is listed and arranged in the following 3 groups:

(low)	(medium)	(high)
3.0–4.9	5.0–6.9	7.0–8.0

The higher the number of types present, the greater the variability. The plot types that occur in over 90% of the sites are then listed in order to give an indication of the relationship between the 2 classifications.

1.1.5 **Species numbers** The species recorded per site are given first, followed by the total number recorded on all the sites of the type. The appropriate categories are as follows:

	(low)	(medium)	(high)
Mean number of species:	60–93	94–126	127–160
Total number of species:	116–170	171–225	226–279

1.2 **Key species of woody perennials**

1.2.1 **Constant trees** Two categories are provided: species occurring in more than 75% of the plots of the original survey are listed without brackets; those occurring in 20–75% of the plots are shown in brackets. Thus, in SITE TYPE 3 (Page 47):

Oak (ash, alder)

indicates that oak occurred in at least 75% of the plots within the sites, and ash and alder each occurred in at least 20% of the plots within the sites but not more than 74%.

1.2.2 **Constant saplings** The saplings are treated as for constant trees, remembering that the breast height diameter of saplings is, by definition, less than 5 cm.

1.2.3 **Constant shrubs** The shrubs are treated as for constant trees, accepting that they are woody perennials that usually do not contribute to the canopy: they are members of the understorey, eg *Sambucus nigra* (elder), *Corylus avellana* (hazel) and *Ilex aquifolium* (holly).

1.2.4 **Trees (basal area)** This entry is an arbiter included to identify large trees. Where a species is listed, the basal area of the trees of that species was at least 0.10 m^2 within a 200 m^2 area of the plots in the site.

2. **Environment**

2.1 **Geographical distribution**
Britain has been divided into 8 areas:
SW   south-west England

17

SE   south-east England
ME   Midlands and East Anglia
NW   northern England, west of the Pennine watershed
NE   northern England, east of the Pennine watershed
Wa   north and south Wales
WS   west and south Scotland
ES   east Scotland

Regions which contained over 30% of the examples of a given site type, as recorded in the original survey, are listed without brackets; regions within brackets have less than 30% of the examples of a given site type and are ranked in order of frequency.

2.2 Solid geology

Geological information was obtained from the 10 inch (1:625 000) Ordnance Survey Geological Map. Each site was assigned to a geological series. Geological series associated with more than 20% of the sites of a particular site type are indicated without brackets; less frequent series are listed within brackets and in diminishing order of frequency.

To reduce numbers of geological series to a manageable size, some were amalgamated:

Code	Abbreviated description	Actual description
A	Calc clay	Calcareous clays and Oxford clays
B	K marl/Lias	Keuper marls, all Lias series, Kimmeridge clay
C	Wealden	Hastings beds, Oldhaven, London clay, Wealden
D	Devonian	Devonian series
E	Oolite/Chalk	Corallian, Cornbrash, Chalk and Southern oolites
F	Carb li/Mag li	Carboniferous and Magnesian limestone
G	Mill grit/Coal mea	Northern oolites, all Coal measures, Millstone grits
H	Silur/Ordov	Silurian and Ordovician series
I	Red s st	Red sandstone series and other sandstone
K	Ign/Metam	Residual igneous and metamorphic types

2.3 Altitude (m)

The mean altitude, in metres, of each replicate of a plot type was calculated from data on 2½ inch (1:25 000) Ordnance Survey maps. The average altitudes were divided into 3 zones:

(low)	(medium)	(high)
58–115	116–172	173–230

2.4 Altitude (bot) and (top)

The mean altitude of the bottom and top altitudes of the woods in the types was derived from 2½ inch (1:25 000) Ordnance Survey maps. Both averages were divided into 3 zones:

(low)	(medium)	(high)
36–68	69–101	102–132
103–203	204–304	305–404

2.5 Slope (0°)

During the original field survey, the slope of the plots was measured in degrees using a Blume-Leiss clinometer. Three categories were recognized:

(low)	(medium)	(high)
6.6–16.5	16.6–26.4	26.5–36.3

2.6 Rainfall (cm)

The average annual rainfall for each plot was taken from the *Climatological atlas of the British Isles* (1952). As with altitude and slope, means were calculated and categorized as follows:

(low)	(medium)	(high)
71–107	108–143	144–180

2.7 Soil

Soil samples from the top 10 cm were taken from the centre of each plot, and pH measurements were made with a glass electrode pH meter. Measurements were taken as soon after collection as possible, suspending soil in distilled water. Means were calculated and arranged in 3 groups:

(low)	(medium)	(high)
4.2–5.0	5.1–5.8	5.9–6.7

2.8 LOI (percentage loss on ignition)

LOI was determined from air-dried soil heated to 450° in a muffle furnace. Means were calculated and arranged in the following groups:

(low)	(medium)	(high)
12.6–21.6	21.7–30.6	30.7–39.6

ACKNOWLEDGEMENTS

M W Shaw was joint leader of the project and was instrumental in the design of the field procedure. He also carried out the computer analysis, and participated throughout in the development of the approach.

Many people have made valuable contributions at different stages of this project, and in particular the following permanent assistants at ITE Merlewood – C J Barr, Mrs Wendy Bowen, Mrs Carole Helliwell and Mrs Judith Johnson. A H F Brown, A D Horrill and J M Sykes, also at Merlewood, contributed valuable advice and assistance with field work in 1971, and M O Hill at ITE Bangor gave statistical advice and analysed some of the data in the Computing Laboratory of the University College of

North Wales, Bangor, whose co-operation is also acknowledged. The following temporary staff also participated, largely in the collection of field data: N E Barber, P A Bassett, S Daggitt, J C Holmes, I P Howes, T J Moss, G K A Reynolds, P L Rye, Mrs C Sargent, Miss C Smith, D J Taylor and P Wilkins. More recently, Professor F T Last has guided the booklet through its final stages. The cover illustration was drawn by C B Benefield.

Our thanks also go to the many landowners and tenants who provided access to their woodlands, often offering hospitality and interest in the project.

REFERENCES

Austin, M. P. & Margules, C. R. 1984. The concept of representativeness in conservation evaluation with particular reference to Australia. (Technical memorandum 84/11). Canberra: CSIRO Institute of Biological Resources, Division of Water and Land Resources.

Birse, E. L. & Robertson, J. S. 1976. Plant communities and soils of the lowland and southern upland regions of Scotland. Aberdeen: Macaulay Institute for Soil Research.

Braun-Blanquet, J. & Tüxen, R. 1952. Irische Pflanzengesellschaften. Die Pflanzenwelt Irlands. Ergebnisse der 9 I.P.E. durch Irland 1949. Veröff. geobot. Inst., Zürich, **25**, 224–420.

Bunce, R. G. H. 1977. The range of variation within the pinewoods. In: Native pinewoods of Scotland, edited by R. G. H. Bunce & J. N. R. Jeffers, 78–87. Cambridge: Institute of Terrestrial Ecology.

Bunce, R. G. H. 1981. The scientific basis of evaluation. In: Values and evaluation, edited by C. I. Rose, 22–27. (Discussion paper in conservation no. 36). London: University College.

Bunce, R. G. H. & Shaw, M. W. 1973. A standardised procedure for ecological survey. J. environ. Manage., **1**, 239–258.

Hill, M. O., Bunce, R. G. H. & Shaw, M. W. 1975. Indicator species analysis: a divisive polythetic method of classification and its application to a survey of native pinewoods in Scotland. J. Ecol., **63**, 597–613.

Kiellund-Lund, J. 1973. A classification of Scandinavian forest vegetation for mapping purposes. IBP i Norden, **11**, 173–206.

Klötzli, F. 1970. Eichen-, Edellaub and Buchwälder der Britischen Inseln. Schweiz. Z. Forstwes., **121**, 329–366.

Koch, W. 1926. Die Vegetationseinheiten der Linthebene. Jb. St Gall. naturw. Ges., **61**, 219–225.

Margules, C. R. & Usher, M. B. 1984. Conservation evaluation in practice. I. Sites of different habitats in north-east Yorkshire, Great Britain. J. environ. Manage., **18**, 153–168.

Peterken, G. F. 1981. Woodland conservation and management. London: Chapman & Hall.

Pilling, R., Gibson, R. A. & Crawley, R. V. 1979. A detailed survey and ecological evaluation of some of the broadleaved woods of the Yorkshire Dales National Park. Bainbridge: Yorkshire Dales National Park Committee.

Ratcliffe, D. A. 1977. Nature conservation review. Cambridge: Cambridge University Press.

Schwickerath, M. 1937. III. Jber. Gruppe Preussen-Rheinl. dtsch. Forstver. Berlin.

Seibert, P. 1969. Uber das Aceri-Fraxinetum als vikariierende Gesellschaft des Galiô-Carpinetum am Rande der Bayerischen Alpen. Vegetatio, **17**, 165–175.

Tittensor, R. 1981. A sideways look at nature conservation in Britain. (Discussion paper no. 29). London: University College.

Tüxen, R. 1937. Die Pflanzengesellschaften Nordwestdeutschlands. Mitt. flor-soz. ArbGemein., **3**, 1–170.

Tüxen, R. 1951. Eindrücke während der Pflanzengeographischen Exkursion durch Süd-Schweden. Vegetatio, **3**, 149–172.

Tüxen, R. 1955. Das system der nordwestdeutschen Pflanzengesellschaften. Mitt. flor-soz. ArbGemein., n.s. **5**, 155–176.

THE KEY

INDICATOR SPECIES

Negative	*Positive*	
	Agrostis canina	1+
	(brown bent-grass)	
	Agrostis tenuis	5+
	(common bent-grass)	
	Anthoxanthum odoratum	1+
	(sweet vernal-grass)	
	Deschampsia flexuosa	5+
	(wavy hair-grass)	
	Galium saxatile	1+
	(heath bedstraw)	
	Sorbus aucuparia	5+
	(rowan)	
	Vaccinium myrtillus	1+
	(bilberry)	
	Dicranum scoparium	1+
	Plagiothecium undulatum	1+
	Polytrichum spp.	5+

Score 4 or less, go to STEP TWO
Score 5 or more, go to STEP NINE

INDICATOR SPECIES

Negative		*Positive*	
BLECHNUM SPICANT	1+	Mercurialis perennis	13+
(HARD-FERN)		(dog's mercury)	
DIGITALIS PURPUREA	1+	Urtica dioica	9+
(FOXGLOVE)		(stinging nettle)	
OXALIS ACETOSELLA	5+		
(WOOD-SORREL)		Thamnium alopecurum	1+
PTERIDIUM AQUILINUM	5+		
(BRACKEN)			
SORBUS AUCUPARIA	1+		
(ROWAN)			
PELLIA SPP.	1+		
POLYTRICHUM SPP.	1+		

Score −1 or less, go to STEP THREE
Score 0 or more, go to STEP SIX

INDICATOR SPECIES

Negative   *Positive*

Arrhenatherum elatius (oatgrass)	1+
Athyrium filix-femina (lady-fern)	5+
Festuca rubra (creeping fescue)	1+
Filipendula ulmaria (meadow-sweet)	5+
Geum urbanum (herb bennet)	5+
Heracleum sphondylium (hogweed)	1+
Mercurialis perennis (dog's mercury)	9+
Rumex acetosa (sorrel)	1+
Taraxacum spp. (dandelion)	1+
Plagiochila asplenioides	1+

Score 4 or less, go to STEP FOUR
Score 5 or more, go to STEP FIVE

INDICATOR SPECIES

Negative		*Positive*	
FAGUS SYLVATICA (BEECH)	5+	*Athyrium filix-femina* (lady-fern)	5+
TAXUS BACCATA (YEW)	1+	*Cardamine flexuosa* (wood bitter-cress)	1+
		Chrysosplenium oppositifolium (opposite-leaved golden saxifrage)	1+
		Dryopteris dilatata (broad buckler-fern)	9+
		Eurhynchium praelongum	9+
		Plagiochila asplenioides	1+
		Plagiothecium denticulatum	1+
		Thuidium tamariscinum	1+
Score 2 or less,	*TYPE 1*	*PTERIDIUM AQUILINUM/QUERCUS-FAGUS* (BRACKEN/OAK-BEECH)	
Score 3 or more,	*TYPE 2*	*HEDERA HELIX/QUERCUS-FRAXINUS* (IVY/OAK-ASH)	

INDICATOR SPECIES

Negative		Positive	
ALNUS GLUTINOSA	5+	*Brachypodium sylvaticum*	5+
(ALDER)		(slender false-brome)	
CHRYSOSPLENIUM		*Galium odoratum*	1+
OPPOSITIFOLIUM	5+	(sweet woodruff)	
(OPPOSITE-LEAVED GOLDEN		*Melica uniflora*	1+
SAXIFRAGE)		(wood melick)	
DIGITALIS PURPUREA	1+	*Polystichum aculeatum*	1+
(FOXGLOVE)		(hard shield-fern)	
GLYCERIA FLUITANS	1+	*Pteridium aquilinum*	5+
(FLOTE GRASS)		(bracken)	
RUMEX CONGLOMERATUS	1+		
(SHARP DOCK)			

Score −2 or less,	*TYPE 3*	*SILENE DIOICA/QUERCUS-FRAXINUS*	
		(RED CAMPION/OAK-ASH)	
Score −1 or more,	*TYPE 4*	*BRACHYPODIUM SYLVATICUM/*	
		QUERCUS-FRAXINUS	
		(SLENDER FALSE-BROME/OAK-ASH)	

25

INDICATOR SPECIES

Negative		*Positive*	
ANGELICA SYLVESTRIS (WILD ANGELICA)	1+	*Anthriscus sylvestris* (cow parsley)	1+
ANEMONE NEMOROSA (WOOD ANEMONE)	1+		
BETULA SPP. (BIRCH)	1+		
DESCHAMPSIA CESPITOSA (TUFTED HAIR-GRASS)	5+		
DRYOPTERIS FILIX-MAS (MALE FERN)	5+		
LONICERA PERICLYMENUM (HONEYSUCKLE)	5+		
RANUNCULUS ACRIS (MEADOW BUTTERCUP)	1+		
LOPHOCOLEA BIDENTATA	1+		
THUIDIUM TAMARISCINUM	1+		

Score −4 or less, go to STEP SEVEN
Score −3 or more, go to STEP EIGHT

INDICATOR SPECIES

Negative		*Positive*	
ACER CAMPESTRE (FIELD MAPLE)	1+	*Acer pseudoplatanus* (sycamore)	5+
AJUGA REPTANS (BUGLE)	5+	*Fagus sylvatica* (beech)	1+
DACTYLORCHIS FUCHSII (COMMON SPOTTED ORCHID)	1+	*Festuca rubra* (creeping fescue)	1+
RUMEX CONGLOMERATUS (SHARP DOCK)	1+	*Poa annua* (annual poa)	1+
		Rumex obtusifolius (broad-leaved dock)	1+
		Hypnum cupressiforme	1+

Score 0 or less,   *TYPE 5*   *GALEOBDOLON LUTEUM/ FRAXINUS-QUERCUS* (YELLOW ARCHANGEL/ASH-OAK)

Score 1 or more,   *TYPE 6*   *MERCURIALIS PERENNIS/FRAXINUS- QUERCUS* (DOG'S MERCURY/ASH-OAK)

27

INDICATOR SPECIES

Negative		*Positive*	
ACER CAMPESTRE	5+	*Chamaenerion angustifolium*	1+
(FIELD MAPLE)		(rosebay willow-herb)	
ARUM MACULATUM	5+	*Cirsium vulgare*	1+
(LORDS-AND-LADIES)		(spear thistle)	
EUONYMUS EUROPAEUS	1+	*Fagus sylvatica*	1+
(SPINDLE-TREE)		(beech)	
	5+	*Holcus lanatus*	1+
EURHYNCHIUM STRIATUM	5+	(Yorkshire fog)	
FISSIDENS TAXIFOLIUS		*Poa trivialis*	13+
		(rough meadow-grass)	

Score −1 or less,	*TYPE 7*	*GALIUM APARINE/ULMUS PROCERA-*	
		FRAXINUS	
		(GOOSEGRASS/ENGLISH ELM-ASH)	
Score 0 or more,	*TYPE 8*	*URTICA DIOICA/FRAXINUS-QUERCUS*	
		(STINGING NETTLE/ASH-OAK)	

INDICATOR SPECIES

Negative	*Positive*	
	Agrostis canina	5+
	(brown bent-grass)	
	Anthoxanthum odoratum	9+
	(sweet vernal-grass)	
	Calluna vulgaris	5+
	(ling)	
	Galium saxatile	9+
	(heath bedstraw)	
	Luzula multiflora	5+
	(many-headed woodrush)	
	Nardus stricta	1+
	(mat-grass)	
	Polygala serpyllifolia	1+
	(common milkwort)	
	Potentilla erecta	9+
	(common tormentil)	
	Hylocomium splendens	5+
	Pseudoscleropodium purum	5+

Score 4 or less, go to STEP TEN
Score 5 or more, go to STEP THIRTEEN

INDICATOR SPECIES

Negative	*Positive*	
	Angelica sylvestris (wild angelica)	1+
	Athyrium filix-femina (lady-fern)	5+
	Filipendula ulmaria (meadow-sweet)	1+
	Ranunculus repens (creeping buttercup)	5+
	Taraxacum spp. (dandelion)	1+
	Valeriana officinalis (valerian)	1+
	Viola riviniana (common violet)	5+
	Mnium undulatum	1+
	Pellia spp.	1+
	Rhytidiadelphus squarrosus	1+

Score 6 or less, go to STEP ELEVEN
Score 7 or more, go to STEP TWELVE

INDICATOR SPECIES

Negative		*Positive*	
CHAMAENERION ANGUSTIFOLIUM (ROSEBAY WILLOW-HERB)	1+	*Dryopteris filix-mas* (male fern)	9+
PINUS SPP. (PINE)	1+	*Fraxinus excelsior* (ash)	5+
		Polypodium vulgare (polypody)	1+
		Dicranum scoparium	5+
		Plagiochila asplenioides	1+
		Plagiothecium undulatum	
		Polytrichum spp.	5+
		Rhytidiadelphus squarrosus	1+
Score 2 or less,	TYPE 9	*PTERIDIUM AQUILINUM/QUERCUS-FAGUS* (BRACKEN/OAK-BEECH)	
Score 3 or more,	TYPE 10	*TEUCRIUM SCORODONIA/QUERCUS-BETULA* (WOOD SAGE/OAK-BIRCH)	

INDICATOR SPECIES

Negative		Positive	
ACER PSEUDOPLATANUS (SYCAMORE)	5+	*Agrostis tenuis* (common bent-grass)	13+
CHAMAENERION ANGUSTIFOLIUM (ROSEBAY WILLOW-HERB)	1+	*Anthoxanthum odoratum* (sweet vernal-grass)	9+
LUZULA SYLVATICA (GREATER WOODRUSH)	+1	*Carex remota* (remote sedge)	5+
MERCURIALIS PERENNIS (DOG'S MERCURY)	1+	*Mentha aquatica* (water mint)	1+
		Polygonum hydropiper	1+
EURHYNCHIUM STRIATUM	5+	(water-pepper)	

Score 0 or less,   *TYPE 11*   *ATHYRIUM FILIX-FEMINA/QUERCUS-FRAXINUS*
(LADY-FERN/OAK-ASH)

Score 1 or more,   *TYPE 12*   *RANUNCULUS REPENS/QUERCUS-ALNUS*
(CREEPING BUTTERCUP/ OAK-ALDER)

INDICATOR SPECIES

Negative		Positive	
CALLUNA VULGARIS (LING)	13+	*Athyrium filix-femina* (lady-fern)	5+
ERIOPHORUM VAGINATUM (COTTON-GRASS)	1+	*Corylus avellana* (hazel)	1+
		Galium aparine (goosegrass)	1+
HYPNUM CUPRESSIFORME	13+	*Geranium robertianum* (herb robert)	1+
		Stellaria holostea (greater stitchwort)	1+
		Eurhynchium praelongum	1+
		Plagiothecium denticulatum	1+

Score 0 or less,	*TYPE 13*	*POTENTILLA ERECTA/ QUERCUS-PINUS-BETULA* (COMMON TORMENTIL/OAK-PINE-BIRCH)	

Score 1 or more, go to STEP FIFTEEN

INDICATOR SPECIES

Negative		Positive	
QUERCUS SPP. (OAK)	5+	*Achillea ptarmica* (sneezewort)	1+
		Betula spp. (birch)	13+
		Epilobium montanum (broad-leaved willow-herb)	5+
		Lathyrus montanus (bitter vetch)	1+
		Rumex obtusifolius (broad-leaved dock)	1+
		Succisa pratensis (devil's-bit scabious)	5+
		Tussilago farfara (coltsfoot)	1+
		Veronica chamaedrys (germander speedwell)	5+
		Rhytidiadelphus triquetrus	5+

Score 4 or less,	*TYPE 14*	*ANTHOXANTHUM ODORATUM/* *QUERCUS* (SWEET VERNAL-GRASS/OAK)
Score 5 or more,	*TYPE 15*	*SUCCISA PRATENSIS/BETULA* (DEVIL'S-BIT SCABIOUS/BIRCH)

34

DETAILED DESCRIPTION OF TYPES OF WOODS

SITE TYPE 1

PTERIDIUM AQUILINUM/QUERCUS-FAGUS (BRACKEN/OAK-BEECH) TYPE

A relatively uniform type most closely related to types 9 and 6, the former being rather more acidic whereas the latter has a higher nutrient status. Plot type 17 [*Pteridium aquilinum/Rubus fruticosus* (bracken/bramble)] covers extensive areas through the woods with localized areas of 9 [*Endymion non-scriptus/Rubus fruticosus* (bluebell/bramble)] and 20 [*Chamaenerion angustifolium/Rubus fruticosus* (rosebay willow-herb/bramble)], but with other plot types scattered throughout. The canopy is principally of oak, beech and birch, with ash, sycamore and rowan locally important, although some other species are occasionally present. The canopy is usually dense although rides and glades are often present. Widespread regeneration of sycamore, birch, beech and ash is taking place. The soils are mainly rather freely drained acidic brown earths, but moister brown earths are locally present in enriched areas, particularly by small streams. Relatively few habitats are present, although there is tall-herb vegetation in the glades. Deer are sometimes present but grazing by domestic stock is unusual because of the surrounding land use. Pheasants are common and evidence of shooting is widespread.

The woods usually occupy gentle slopes or are on level ground, and are set in lowland landscapes covering a small altitudinal range. They are mainly surrounded by arable land, although there are some leys and other woodlands often nearby. The sites are mainly old coppices, now usually neglected, although some conifers may have been planted, and there has been some conversion to high forest.

Although likely to be centred on the Weald and south-east England, this type also extends elsewhere in southern and south-west England, as well as into the Welsh borders and the Forest of Dean. It is less common elsewhere in the lowlands of the midlands and northern England.

Within the *Nature conservation review*, sites such as Blean Woods (south-east England), the Mens and the Cut (south-east England) and Hamstreet (south-east England) are likely to be included in this type.

On the Continent, comparable sites, although usually with a higher proportion of beech in the canopy, probably occur quite commonly in northern Germany, Belgium, Holland and the lowlands of France.

The sites in this type would probably be called, in general terms, oak-beech woods over bracken and bramble on acidic lowland clays. The main phytosociological associations present in the woods are probably Fago-Quercetum petraeae Tx 1955, Blechno-Quercetum Br-Bl et Tx 1952 and Betulo-Quercetum Tx 1937.

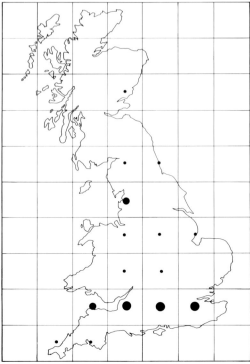

A wood occupying a gentle sloping site in southern England. The canopy comprises *Quercus* spp. (oak), *Fraxinus excelsior* (ash) and *Fagus sylvatica* (beech) with some emergent conifers.

Type 1. *Pteridium aquilinum/Quercus-Fagus* (bracken/oak-beech)

17.   *Pteridium aquilinum/Rubus fruticosus* (bracken/bramble) type

A type of low heterogeneity with *Ilex aquifolium* (holly), *Fagus sylvatica* (beech) and *Carpinus betulus* (hornbeam) as selective species. The canopy is usually dense consisting of oak and beech, but there are few saplings of shrubs. *Rubus fruticosus* (bramble), *Pteridium aquilinum* (bracken) and *Hedera helix* (ivy) are the usual ground cover species. The type would probably be termed dry, acid sessile oak woodland and the soils are mainly brown earths.

9.   *Endymion non-scriptus/Rubus fruticosus* (bluebell/bramble) type

A type of medium heterogeneity with *Fagus sylvatica* (beech), *Acer pseudoplatanus* (sycamore) and *Endymion non-scriptus* (bluebell) as selective species. The canopy is usually dense consisting of oak, sycamore and beech with saplings and an understorey of hazel sometimes present. *Rubus fruticosus* (bramble), *Pteridium aquilinum* (bracken) and *Dryopteris dilatata* (broad buckler-fern), are the usual ground cover species. The type would probably be called mixed deciduous woodland and the soils are usually brown earths with a tendency towards a deposition of iron.

20.   *Chamaenerion angustifolium/Rubus fruticosus* (rosebay willow-herb/bramble) type

A type of medium heterogeneity with *Carpinus betulus* (hornbeam), *Holcus lanatus* (Yorkshire fog) and *Juncus effusus* (soft-rush) as selective species. The canopy is of average density with few saplings or shrubs. *Rubus fruticosus* (bramble), *Pteridium aquilinum* (bracken) and *Deschampsia flexuosa* (wavy hair-grass) are the usual ground cover species. The type would probably be termed mixed sessile oak-birch woodland; the soils are mainly acid brown earth and are often rather heavy.

1.   *Urtica dioica/Rubus fruticosus* (stinging nettle/bramble) type

A type of low heterogeneity with *Sambucus nigra* (elder), *Mercurialis perennis* (dog's mercury) and *Euonymus europaeus* (spindle-tree) as selective species. The canopy is usually dense consisting of oak and beech with saplings often present, as well as an understorey of hazel. *Rubus fruticosus* (bramble), *Mercurialis perennis* (dog's mercury) and *Hedera helix* (ivy) are the usual ground cover species. The type would probably be referred to as mixed deciduous woodland and the soils are mainly eutrophic brown earths.

VEGETATION

Key species
Constant species: *Rubus fruticosus* (bramble), *Pteridium aquilinum* (bracken), *Quercus* spp. (oak), *Fagus sylvatica* (beech)

Plot dominants: *Rubus fruticosus* (bramble), *Pteridium aquilinum* (bracken), *Hedera helix* (ivy), *Lonicera periclymenum* (honeysuckle)

Selective species: *Fagus sylvatica* (beech), *Ilex aquifolium* (holly), *Chamaenerion angustifolium* (rosebay willow-herb), *Pteridium aquilinum* (bracken), *Rubus fruticosus* (bramble), *Dicranella heteromalla*

Blend of   Frequency 17, 9, 20 (1, 11, 18, 19, 23, 24)
plot types: · Mean number 6.1 (med)

Mean number of species: 78 (low)
Total number of species: 196 (med)

Canopy and understorey species

Constant trees	*Constant saplings*
Oak (beech)	(Sycamore, birch, beech)

Constant shrubs	*Trees (basal area)/plot over 0.1 m^2*
(Hazel) (low density)	Oak, beech (average canopy)

ENVIRONMENT

Geographical distribution	*Solid geology*	*Rainfall (cm)*
SE (S)	Wealden, Oolite/Chalk	76 *(low)*

Altitude (m)	*Altitude (bot)*	*Altitude (top)*	*Slope (°)*	*Soil (pH)*	*LOI*
106 *(low)*	73 *(med)*	138 *(low)*	12.5 *(low)*	4.2 *(low)*	12.6 *(low)*

SITE TYPE 2

HEDERA HELIX/QUERCUS-FRAXINUS (IVY/OAK-ASH) TYPE

A quite variable type most closely related to types 1 and 9, the former being more uniformly acidic throughout and the latter having areas of vegetation associated with more strongly acid conditions. Plot types 10 [*Athyrium filix-femina/Rubus fruticosus* (lady-fern/bramble)] and 11 [*Potentilla sterilis/Rubus fruticosus* (barren strawberry/bramble)] occur in quite large areas through most of the woods, but the other plot types are scattered throughout according to local variations in topography. The canopy is usually a mixture of oak, ash and birch, with sycamore, rowan and willow locally important. Regeneration of ash, birch, sycamore and hawthorn is common.

The soils are principally brown earths, but there are usually extensive areas of acid brown earths. Gleys and gleyed brown earths are present in limited areas. Relatively few non-woodland habitats are present, although the majority of woods have small streams, with associated marshy areas, running through them. Rides and glades are widespread, many of the woods having evidence of management for pheasants.

The woods are usually on more or less level land covering a small altitudinal range, but are occasionally on somewhat steeper slopes. The surrounding farmland is usually intensively managed and, although arable is the most common land use, short-term grassland is also frequent.

The type occurs widely throughout southern and south-west England, but also extends to south Wales and the south midlands. There are probably other sites present in the lowlands of south-west England and north Wales.

Within the *Nature conservation review*, sites such as Blaenau Nedd and Mellte (south Wales), Asham Wood (south-west England) and Wychwood (south England) are likely to be included in this type.

On the Continent, comparable sites will occur in northern Germany, Belgium and the French lowlands, decreasing in frequency away from the Atlantic fringe. However, the majority of species present are in Continental woods and there is much in common with such vegetation.

The sites in this type would probably be called, in general terms, acid lowland woods, with localized enrichment, or oak-ash woods over bramble and bracken. The main phytosociological associations present in the woods are probably Ulmo-Fraxinetum E Sjogren ap KL 1973 (Ulmo-Quercetum Tx 1931), Dryopteride-Fraxinetum Klötzli 1970 and Fago-Quercetum petraeae Tx 1955.

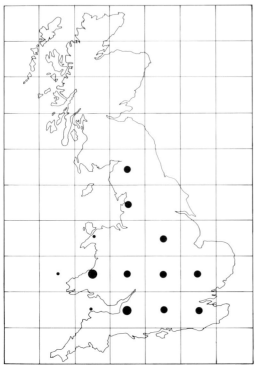

A wood occupying level ground in the midlands of England. The canopy is mainly of *Fraxinus excelsior* (ash) and *Quercus* spp. (oak), but *Acer pseudoplatanus* (sycamore) and *Betula* spp. (birch) are also present. Some dead *Ulmus procera* (English elm) may be seen on the edge of the wood.

Type 2. *Hedera helix/ Quercus-Fraxinus* (ivy/oak-ash)

10.   *Athyrium filix-femina/Rubus fruticosus* (lady-fern/bramble) type
A type of medium heterogeneity with *Circaea lutetiana* (enchanter's nightshade),
Veronica montana (wood speedwell) and *Dryopteris dilatata* (broad buckler-fern) as
selective species. The ground cover usually consists mainly of *Rubus fruticosus*
(bramble). The canopy is dense, mainly of oak and ash but also sycamore, birch,
willow and alder, with few saplings and an understorey of hazel often present. The
type would probably be termed moist pedunculate oak-ash woodland. The soils are
mainly brown earths, although often rather rocky and sometimes shallow.

11.   *Potentilla sterilis/Rubus fruticosus* (barren strawberry/bramble) type
A type of medium heterogeneity, with *Viola riviniana* (common violet), *Fragaria vesca*
(wild strawberry) and *Prunella vulgaris* (self heal) as selective species. The canopy is
quite dense usually of oak and ash but also with birch and willow, and there is an
understorey of hazel. The ground vegetation is invariably dominated by *Rubus
fruticosus* (bramble). The type would probably be included as pedunculate oak
woodland or mixed deciduous woodland on lower valley sides. The soils are mainly
brown earths.

5.   *Glechoma hederacea/Mercurialis perennis* (ground ivy/dog's mercury) type
A type of medium heterogeneity with *Geum urbanum* (herb bennet), *Brachypodium
sylvaticum* (slender false-brome) and *Urtica dioica* (stinging nettle) as selective
species. The canopy is of average density with ash and oak as the main species, and
an understorey of hazel is usually present. *Mercurialis perennis* (dog's mercury) and
Rubus fruticosus (bramble) are the usual ground cover species. The type would
probably be called mixed deciduous woodland and the soils are mainly calcareous
brown earths, with gleying sometimes present.

7.   *Carex sylvatica/Rubus fruticosus* (wood sedge/bramble) type
A type of medium heterogeneity with *Acer campestre* (field maple), *Sorbus
torminalis* (wild service tree) and *Euphorbia amygdaloides* (wood spurge) as selective
species. The ground cover usually consists of *Rubus fruticosus* (bramble) and
Mercurialis perennis (dog's mercury). The canopy is of average density consisting
mainly of oak and ash with a dense understorey of hazel present in most areas. The
type would probably be called mixed deciduous woodland. The soils are usually
brown earths, although there is a tendency for gleying to take place.

SUMMARY OF **SITE TYPE 2**
HEDERA HELIX/QUERCUS-FRAXINUS (IVY/OAK-ASH) TYPE

VEGETATION

Key species

Constant species: *Rubus fruticosus* (bramble), *Dryopteris dilatata* (broad buckler-fern), *Quercus* spp. (oak), *Fraxinus excelsior* (ash)

Plot dominants: *Rubus fruticosus* (bramble), *Pteridium aquilinum* (bracken), *Holcus mollis* (creeping soft-grass), *Hedera helix* (ivy)

Selective species: *Hedera helix* (ivy), *Lonicera periclymenum* (honeysuckle), *Dryopteris dilatata* (broad buckler-fern), *Rubus fruticosus* (bramble), *Veronica montana* (wood speedwell), *Circaea lutetiana* (enchanter's nightshade)

Blend of plot types: Frequency 10, 11, (5, 7, 9, 12, 13, 14, 17, 22, 24) Mean number 6.1 (med)

Mean number of species: 100 (med)
Total number of species: 233 (high)

Canopy and understorey species

Constant trees
Oak, ash (birch)

Constant saplings
Ash (birch) (high density)

Constant shrubs
(Hazel) (average density)

Trees (basal area)/plot over 0.1 m²
Oak (ash, birch) (average canopy)

ENVIRONMENT

Geographical distribution	*Solid geology*	*Rainfall (cm)*
Wa, SW (ME, NE)	No dominant (A, B, C, G, H)	99 *(low)*

Altitude (m)	*Altitude (bot)*	*Altitude (top)*	*Slope (°)*	*Soil (pH)*	*LOI*
83 *(low)*	66 m *(low)*	115 m *(low)*	15.8 *(low)*	4.6 *(low)*	12.6 *(low)*

SITE TYPE 3

SILENE DIOICA/QUERCUS-FRAXINUS (RED CAMPION/OAK-ASH) TYPE

A quite variable type most closely related to site types 12 and 15, the former being associated with steeper terrain and the latter generally more upland in nature. However, all the sites are linked by the high frequency of waterlogged conditions. Plot types 14 [*Chrysosplenium oppositifolium/Rubus fruticosus* (opposite-leaved golden saxifrage/bramble)] and 9 [*Endymion non-scriptus/Rubus fruticosus* (bluebell/bramble)] occur most widely in the woods, with a limited range of other types scattered throughout, particularly on drier banks. The canopy is mainly oak and ash, but alder predominates by the watercourses, with sycamore, willow and other species, such as bird cherry, also present. Regeneration of ash, hawthorn and willow often occurs locally. The woods are often rather open and although mainly consisting of coppiced stems, have not been managed regularly as such, exploitation being generally haphazard. Some exotics have often been planted in groups throughout the woods.

The soils are mainly dominated by waterlogged conditions and are invariably gleyed. There is often a humus-rich surface horizon, with a high silt content deposited by the streams. Otherwise, gleyed brown earths are widespread, with brown earths on the better-drained areas. Slow-flowing streams or rivers usually run through the woods and extensive marshy areas are present. Glades are often associated with the wettest areas and have extensive tall-herb vegetation and clumps of brambles.

Deer are sometimes present and many of the woods frequently have sections open to cattle and sheep grazing.

The woods are usually on more or less level land or on shallow river valley sides occupying a small altitudinal range. The surrounding farmland is usually pasture but sometimes arable, and small woods are present.

Although the surveyed sites were mainly in south Wales, this type is likely to have a much wider distribution extending into south-west, north-west and north-east England as well as the Weald, always being linked to similar topographic conditions.

Within the *Nature conservation review*, sites such as the Fal Ruan Estuary (south-west England), Saxonbury Hill (south-east England) and Cwm Clydach (south Wales) are likely to be included in this type.

The high water levels probably mean that the type is likely to be restricted on the Continent to the Atlantic fringe. Comparable areas are therefore likely to be present in Brittany, Belgium and Ireland. Such sites may also be present on the Continent beside big rivers, such as the Rhine, as many of the species are wide-ranging under suitable soil conditions.

The sites in this wood would probably be called, in general terms, ash-alder woods with variable ground vegetation, or marshy woods by small streams with localized drier areas. The main phytosociological associations present in the woods are probably Alno-Fraxinetum KL ap Seibert 1969, Carici-remotae Fraxinetum (W Koch 1926) Schwickerath 1937.

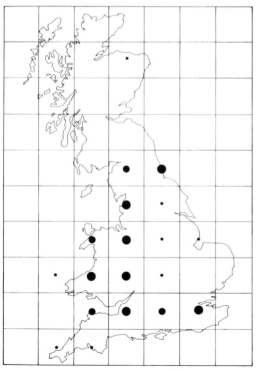

A wood occupying almost level ground in west Wales. The canopy is mainly *Fraxinus excelsior* (ash), *Alnus glutinosa* (alder) and *Acer pseudoplatanus* (sycamore), but *Salix* spp. (willow) is also present.

Type 3. *Silene dioica/ Quercus-Fraxinus* (red campion/oak-ash)

14.   *Chrysosplenium oppositifolium/Rubus fruticosus* (opposite-leaved golden saxifrage/bramble) type
A type of medium heterogeneity and an average species complement, with *Iris pseudacorus* (yellow flag), *Galium palustre* (marsh bedstraw) and *Solanum dulcamara* (bittersweet) as selective species. The canopy is usually dense with ash and alder as the major species, although willow and birch are widespread. An understorey is usually present, with many saplings. The ground cover is principally *Rubus fruticosus* (bramble), but *Chrysosplenium oppositifolium* (opposite-leaved golden saxifrage) and *Urtica dioica* (stinging nettle) are common also. The type is mainly ash-alder woodland growing on eutrophic valley floors. The soils are invariably heavily gleyed, often with a surface humus-rich horizon that is waterlogged.

9.   *Endymion non-scriptus/Rubus fruticosus* (bluebell/bramble) type
A type of medium heterogeneity with *Fagus sylvatica* (beech), *Acer pseudoplatanus* (sycamore) and *Endymion non-scriptus* (bluebell) as selective species. The canopy is usually dense consisting of oak, sycamore and beech with saplings and an understorey of hazel sometimes present. *Rubus fruticosus* (bramble), *Pteridium aquilinum* (bracken) and *Dryopteris dilatata* (broad buckler-fern) are the usual ground cover species. The type would probably be called mixed deciduous woodland and the soils are usually brown earths with a tendency towards a deposition of iron.

10.   *Athyrium filix-femina/Rubus fruticosus* (lady-fern/bramble) type
A type of medium heterogeneity with *Circaea lutetiana* (enchanter's nightshade), *Veronica montana* (wood speedwell) and *Dryopteris dilatata* (broad buckler-fern) as selective species. The ground cover usually consists mainly of *Rubus fruticosus* (bramble). The canopy is dense, mainly of oak and ash but also sycamore, birch, willow and alder, with few saplings and an understorey of hazel often present. The type would probably be termed moist pedunculate oak-ash woodland. The soils are mainly brown earths, although often rather rocky and sometimes shallow.

15.   *Galium palustre/Agrostis tenuis* (marsh bedstraw/common bent-grass) type
A type of medium heterogeneity with a high species complement, *Senecio aquaticus* (marsh ragwort), *Scrophularia aquatica* (water betony) and *Lotus pedunculatus* (large birdsfoot-trefoil) being selective species. The canopy of alder and ash is usually open, with few saplings or shrubs. The ground cover is mainly *Agrostis tenuis* (common bent-grass) but a wide range of other species are present, eg *Holcus lanatus* (Yorkshire fog) and *Rubus fruticosus* (bramble). The type would usually be called scrub or occasionally grassland with scattered trees. The soils are mainly gleyed brown earths but some brown earths are also present.

VEGETATION

Key species

Constant species: *Rubus fruticosus* (bramble), *Dryopteris dilatata* (broad buckler-fern), *Fraxinus excelsior* (ash), *Eurhynchium praelongum*

Plot dominants: *Rubus fruticosus* (bramble), *Holcus mollis* (creeping soft-grass), *Holcus lanatus* (Yorkshire fog), *Dryopteris dilatata* (broad buckler-fern)

Selective species: *Silene dioica* (red campion), *Alnus glutinosa* (alder), *Chrysosplenium oppositifolium* (opposite-leaved golden saxifrage), *Rumex conglomeratus* (sharp dock), *Ranunculus repens* (creeping buttercup), *Athyrium filix-femina* (lady-fern)

Blend of plot types: Frequency 14, 9, 10, 15, 12, (13, 21, 22, 31, 32)
Mean number 6.2 (med)

Mean number of species: 117 (med)
Total number of species: 228 (high)

Canopy and understorey species

Constant trees
Oak (ash, alder)

Constant saplings
Ash (low density)

Constant shrubs
(Hazel) (low density)

Trees (basal area)/plot over 0.1 m^2
Oak (ash, alder, sycamore) (average canopy)

ENVIRONMENT

Geographical distribution	*Solid geology*	*Rainfall (cm)*
Wa, WS	Devonian (B, H, I)	122 *(med)*

Altitude (m)	*Altitude (bot)*	*Altitude (top)*	*Slope (°)*	*Soil (pH)*	*LOI*
66 *(low)*	75 m *(low)*	103 m *(low)*	15.7 *(low)*	5.0 *(med)*	18.9 *(low)*

47

SITE TYPE 4

BRACHYPODIUM SYLVATICUM/QUERCUS-FRAXINUS (SLENDER FALSE-BROME/
OAK-ASH) TYPE

This type is variable and is most closely related to types 2 and 3, being somewhat more calcareous than the former, and better drained than the latter. Plot type 12 [*Geum urbanum/Mercurialis perennis* (herb bennet/dog's mercury)] occurs throughout the woods, but types 5, 10 and 1 are also widespread with a scattering of other types. Ash is the major canopy species, hawthorn and rowan being secondary species, but a wide range of other species are also present. Ash and hawthorn are regenerating widely, with birch, sycamore and holly locally important.

The major soil types are brown earths, with a varying degree of calcium present depending upon parent material. Some acid brown earths are also present, mainly because of deposition of acidic glacial material – in contrast to the southern limestone types 5 and 6, which therefore have less complex vegetation.

The woods are growing mainly on quite steep slopes. Many rocky habitats are present due to the frequent outcropping of the rock, and occupy quite a wide altitudinal range. The sites are often complex topographically and sometimes have small streams running through them at the bottom of the slopes.

There are a limited number of rides and glades, and the woods are often adjacent to other woodland blocks; there is a range of other land uses, particularly arable and pasture. Roads often pass through the woods and the frequent presence of houses nearby indicates a degree of disturbance.

The woods have usually been managed as coppices in the past or locally as high forest, but are now mainly neglected, with only haphazard exploitation. There are some scattered exotics. Deer are often present; stock are in some woods but completely absent elsewhere. There are therefore strong contrasts in grazing pressures.

The geographical distribution of this type is strongly linked to the associated geological series. Thus, it occurs mainly in areas such as the lower Wye valley, north Lancashire and the Yorkshire Dales. It is also likely to be found elsewhere on comparable substrata in areas such as north Wales and the Derbyshire Dales.

Within the *Nature conservation review*, sites such as Whitbarrow (north-west England), Reins Wood (north-east England) and various of the Wye Gorge complex (south Wales) are likely to be included in this type.

On the Continent, this type is likely to be very infrequent, as the degree of deposition of acid material on limestone is not extensive, perhaps only occurring in Belgium, although locally lime-rich areas on otherwise acidic rocks may have the same effect. However, in the west of Ireland, such sites could well be quite common.

In general terms, the sites in this type would probably be called ash-oak woods with complex ground vegetation, or northern limestone woods. The main phytosociological associations present in the woods are probably Ulmo-Fraxinetum E Sjogren ap KL 1973 (Ulmo-Quercetum Tx 1931), Dryopterido-Fraxinetum Klötzli 1970 and Querco-Fraxinetum Klötzli 1970.

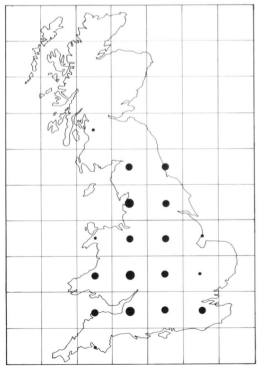

A wood occupying a steep slope in north-east England. The canopy is mainly *Fraxinus excelsior* (ash) and *Betula* spp. (birch), but *Acer pseudoplatanus* (sycamore), *Corylus avellana* (hazel) and *Salix* spp. (willow) are also present.

Type 4. *Brachypodium sylvaticum/Quercus-Fraxinus* (slender false-brome/oak-ash)

49

12.   *Geum urbanum/Mercurialis perennis* (herb bennet/dog's mercury) type
A type of medium heterogeneity and a high species complement, with *Brachypodium sylvaticum* (slender false-brome), *Fragaria vesca* (wild strawberry) and *Sanicula europea* (sanicle) as selective species. The canopy is usually quite dense with oak and ash as major species, but birch, sycamore, alder and hawthorn are also present locally. There is an understorey of hazel, with saplings of low density but frequent occurrence. The ground cover is principally *Mercurialis perennis* (dog's mercury) and *Rubus fruticosus* (bramble). The types belongs to the broad range of pedunculate oak-ash woodland on quite basiphilous valley sides. The soil are mainly brown earths, often skeletal, and usually gleyed.

5.   *Glechoma hederacea/Mercurialis perennis* (ground ivy/dog's mercury) type
A type of medium heterogeneity with *Geum urbanum* (herb bennet), *Brachypodium sylvaticum* (slender false-brome) and *Urtica dioica* (stinging nettle) as selective species. The canopy is of average density with ash and oak as the main species, and an understorey of hazel is usually present. *Mercurialis perennis* (dog's mercury) and *Rubus fruticosus* (bramble) are the usual ground cover species. The type would probably be called mixed deciduous woodland and the soils are mainly calcareous brown earths, with gleying sometimes present.

10.   *Athyrium filix-femina/Rubus fruticosus* (lady-fern/bramble) type
A type of medium heterogeneity with *Circaea lutetiana* (enchanter's nightshade), *Veronica montana* (wood speedwell) and *Dryopteris dilatata* (broad buckler-fern) as selective species. The ground cover usually consists mainly of *Rubus fruticosus* (bramble). The canopy is dense, mainly of oak and ash but also sycamore, birch, willow and alder, with few saplings and an understorey of hazel often present. The type would probably be termed moist pedunculate oak-ash woodland. The soils are mainly brown earths, although often rather rocky and sometimes shallow.

11.   *Potentilla sterilis/Rubus fruticosus* (barren strawberry/bramble) type
A type of medium heterogeneity, with *Viola riviniana* (common violet), *Fragaria vesca* (wild strawberry) and *Prunella vulgaris* (self heal) as selective species. The canopy is quite dense usually of oak and ash but also with birch and willow, and there is an understorey of hazel. The ground vegetation is invariably dominated by *Rubus fruticosus* (bramble). The type would probably be included as pedunculate oak woodland or mixed deciduous woodland on lower valley sides. The soils are mainly brown earths.

SUMMARY OF **SITE TYPE 4**

BRACHYPODIUM SYLVATICUM/QUERCUS-FRAXINUS (SLENDER FALSE-BROME/ OAK-ASH) TYPE

VEGETATION

Key species

Constant species:	*Fraxinus excelsior* (ash), *Rubus fruticosus* (bramble), *Dryopteris filix-mas* (male fern), *Corylus avellana* (hazel)
Plot dominants:	*Mercurialis perennis* (dog's mercury), *Rubus fruticosus* (bramble), *Luzula sylvatica* (greater woodrush), *Deschampsia cespitosa* (tufted hair-grass)
Selective species:	*Allium ursinum* (ramsons), *Brachypodium sylvaticum* (slender false-brome), *Heracleum sphondylium* (hogweed), *Arrhenatherum elatius* (oat-grass), *Mercurialis perennis* (dog's mercury), *Fragaria vesca* (wild strawberry)
Blend of plot types:	Frequency 12, 5, 10, 11 (7, 8, 9, 22, 31) Mean number 6.4 (med)

Mean number of species: 126 (med)
Total number of species: 277 (high)

Canopy and understorey species

Constant trees	*Constant saplings*
Ash (oak, hawthorn, birch)	Ash (hawthorn) (average density)
Constant shrubs	*Trees (basal area)/plot over 0.1 m^2*
(Hazel) (average density)	Oak (ash) (open canopy)

ENVIRONMENT

Geographical distribution	Solid geology	Rainfall (cm)
NE (NW, Wa, SW)	Carb li/Mg li, Mill grit/ Coal mea (E, D)	91 *(low)*

Altitude (m)	Altitude (bot)	Altitude (top)	Slope (°)	Soil (pH)	LOI
108 *(low)*	84 m *(med)*	163 m *(low)*	27 *(high)*	5.5 *(med)*	13.8 *(low)*

SITE TYPE 5

GALEOBDOLON LUTEUM/FRAXINUS-QUERCUS (YELLOW ARCHANGEL/ASH-OAK) TYPE

A relatively uniform type most closely related to types 6 and 7, having more southern species than the former and less heavy soils than the latter. Plot types 5 [*Glechoma hederacea/Mercurialis perennis* (ground ivy/dog's mercury)] and 7 [*Carex sylvatica/ Rubus fruticosus* (wood sedge/bramble)] predominate in extensive stands, with a range of other plot types scattered throughout. Ash is consistently the major canopy species, although oak is also widespread. Otherwise, hawthorn, field maple and birch occur widely, with other species occasionally present. A dense shrub layer is often present limiting regeneration. The canopy is usually of ash but hawthorn and sycamore and a range of other minor species are also locally abundant in more open areas.

Calcareous brown earths are the major soil type present, with rendzinas around the rocky outcrops. Gleying is locally important, particularly where clay overlies the bedrock, and some brown earths are also present.

In general, few additional habitats are present, although the glades and rides tend to have dense and variable vegetation.

Deer are widespread and, although domestic cattle are not usually present, some woods are open to grazing. Other woodland sites are often adjacent, with arable land and pasture predominating nearby land uses. The nearness of housing and roads indicates frequent disturbance. The woods are usually on level or gently sloping sites and occupy typical lowland landscapes, covering a narrow altitudinal range.

The woods are mainly neglected coppices, but there has often been some recent cutting, usually of a dispersed nature. There was no evidence of silvicultural treatment but some sites are returning to high forest and management of pheasant shooting is important in many sites.

The distribution of this type is centred on the lowlands around the margins of the Severn estuary, but extends north into the Welsh borders, east into the midlands and also into southern England. Outliers are also likely in East Anglia, south-west England and the Yorkshire wolds.

Within the *Nature conservation review*, sites such Rodney Stoke (south-west England), Salisbury Wood (south Wales) and Hales Wood (East Anglia) are likely to be included in this type.

The high proportion of species with Continental affinities suggests that such sites are likely to be present widely on the Continent in northern Germany, Holland, Belgium and throughout lowland France. However, away from the Atlantic margins, such sites are likely to have less bramble in the ground vegetation and beech as a more important canopy species than ash. Comparable sites are also likely in Ireland.

The sites in this type would probably be called, in general terms, ash-oak woods over bramble and mercury, or calcareous lowland woods. The main phytosociological associations present in the woods are probably Ulmo-Fraxinetum E Sjogren ap KL 1973, Querco-Fraxinetum Klötzli 1970, and Dryopterido-Fraxinetum Klötzli 1970.

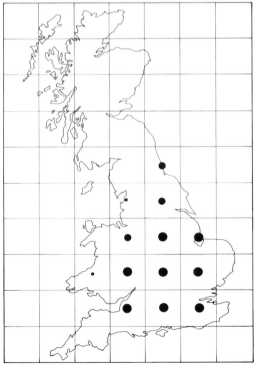

A wood occupying a rounded slope in southern England. The canopy is a mixture of *Fraxinus excelsior* (ash) and *Quercus* spp. (oak), but a variety of other species are also present and a group of *Populus* spp. (poplar) may be seen on the right.

Type 5. *Galeobdolon luteum/Fraxinus-Quercus* (yellow archangel/ash-oak)

5.   *Glechoma hederacea/Mercurialis perennis* (ground ivy/dog's mercury) type
A type of medium heterogeneity with *Geum urbanum* (herb bennet), *Brachypodium sylvaticum* (slender false-brome) and *Urtica dioica* (stinging nettle) as selective species. The canopy is of average density with ash and oak as the main species, and an understorey of hazel is usually present. *Mercurialis perennis* (dog's mercury) and *Rubus fruticosus* (bramble) are the usual ground cover species. The type would probably be called mixed deciduous woodland and the soils are mainly calcareous brown earths, with gleying sometimes present.

7.   *Carex sylvatica/Rubus fruticosus* (wood sedge/bramble) type
A type of medium heterogeneity with *Acer campestre* (field maple), *Sorbus torminalis* (wild service tree) and *Euphorbia amygdaloides* (wood spurge) as selective species. The ground cover usually consists of *Rubus fruticosus* (bramble) and *Mercurialis perennis* (dog's mercury). The canopy is of average density consisting mainly of oak and ash with a dense understorey of hazel present in most areas. The type would probably be called mixed deciduous woodland. The soils are usually brown earths, although there is a tendency for gleying to take place.

10.   *Athyrium filix-femina/Rubus fruticosus* (lady-fern/bramble) type
A type of medium heterogeneity with *Circaea lutetiana* (enchanter's nightshade), *Veronica montana* (wood speedwell) and *Dryopteris dilatata* (broad buckler-fern) as selective species. The ground cover usually consists mainly of *Rubus fruticosus* (bramble). The canopy is dense, mainly of oak and ash but also sycamore, birch, willow and alder, with few saplings and an understorey of hazel often present. The type would probably be termed moist pedunculate oak-ash woodland. The soils are mainly brown earths, although often rather rocky and sometimes shallow.

1.   *Urtica dioica/Rubus fruticosus* (stinging nettle/bramble) type
A type of low heterogeneity with *Sambucus nigra* (elder), *Mercurialis perennis* (dog's mercury) and *Euonymus europaeus* (spindle-tree) as selective species. The canopy is usually dense consisting of oak and beech with saplings often present, as well as an understorey of hazel. *Rubus fruticosus* (bramble), *Mercurialis perennis* (dog's mercury) and *Hedera helix* (ivy) are the usual ground cover species. The type would probably be referred to as mixed deciduous woodland and the soils are mainly eutrophic brown earths.

GALEOBDOLON LUTEUM/FRAXINUS-QUERCUS (YELLOW ARCHANGEL/ASH-OAK) TYPE

VEGETATION

Key species
Constant species: *Fraxinus excelsior* (ash), *Rubus fruticosus* (bramble), *Corylus avellana* (hazel), *Eurhynchium praelongum*

Plot dominants: *Rubus fruticosus* (bramble), *Mercurialis perennis* (dog's mercury), *Urtica dioica* (stinging nettle), *Deschampsia cespitosa* (tufted hair-grass)

Selective species: *Acer campestre* (common maple), *Galeobdolon luteum* (yellow archangel), *Mercurialis perennis* (dog's mercury), *Glechoma hederacea* (ground ivy), *Arum maculatum* (lords-and-ladies), *Circaea lutetiana* (enchanter's nightshade)

Blend of   Frequency 5, 7, 10, (1, 2, 3, 4, 6, 11)
plot types: Mean number 4.6 (low)

Mean number of species: 94 (med)
Total number of species: 220 (med)

Canopy and understorey species
Constant trees   *Constant saplings*
Ash, oak (hawthorn, field maple)   Ash, hawthorn (high density)

Constant shrubs   *Trees (basal area)*
Hazel (elder) (high density)   Ash, oak (average canopy)

ENVIRONMENT

Geographical distribution	*Solid geology*	*Rainfall (cm)*
SW, ME	Oolite/Chalk (A, B, D)	81 *(low)*

Altitude (m)	*Altitude (bot)*	*Altitude (top)*	*Slope (°)*	*Soil (pH)*	*LOI*
113 *(low)*	84 m *(med)*	133 m *(low)*	14.9 *(low)*	5.8 *(med)*	13.4 *(low)*

55

SITE TYPE 6

MERCURIALIS PERENNIS/FRAXINUS-QUERCUS (DOG'S MERCURY/ASH-OAK) TYPE

A variable type most closely related to types 5 and 7, the former being somewhat more uniform in habitats and degree of disturbance whereas the latter type occupies heavier soils in the main. As with the previous type, plot type 5 [*Glechoma hederacea/Mercurialis perennis* (ground ivy/dog's mercury)] is the most abundant throughout the woods, although it is never quite so dominant. Otherwise, types 7 [*Carex sylvatica/Rubus fruticosus* (wood sedge/bramble)] and 12 [*Geum urbanum/ Mercurialis perennis* (herb bennet/dog's mercury)] are particularly persistent, with other types occurring only locally.

The soils are mainly calcareous brown earths but brown earths and calcareous gleys are also present, being derived from a range of calcareous substrata. Relatively few habitats are present, although rock outcrops are common, and aquatic features are rare. Glades and rides are often present with tall-herb vegetation growing vigorously. Deer are sometimes present and rabbits were recorded in all sites, but domestic stock are not usual.

The woods are usually situated on gentle slopes but some steeper areas are sometimes present and, although usually occupying a narrow altitudinal range, sometimes occupy a wider range. The most usual surrounding land use is arable but grassland, usually leys, is almost as common. Although the woods are not often grazed, cattle and sheep sometimes get into the woods and cause extensive disturbance on the soft soil surfaces. Ash is regenerating widely in most sites and sycamore and hawthorn are widespread, with a range of other species also frequently present. The sites have usually been managed as a mixture of coppice and high forest but, in the main, not as intensively as some other types, with much haphazard removal of individual trees.

This type follows the patterns of the associated geology and thus has a widespread distribution throughout lowland England. It is, however, likely to be particularly common on the Mendips and on the chalk formations of the south of England. Further north, sites are present upon carboniferous and magnesian limestone strata in north Lancashire and Durham respectively.

Within the *Nature conservation review*, sites such as Ebbor Gorge (south-west England), Weston Big Wood (south-west England) and Oxwich Point (south Wales) are likely to be included in this type.

On the Continent, similar sites probably occur on calcareous formations in Belgium and northern France, perhaps extending further south and east with progressively less oceanic species. Comparable sites are probably present in Ireland, but they would have more hazel and less ash and oak in the canopy.

The sites in this type would probably be called, in general terms, ash woods over dog's mercury, or freely drained calcareous woods. The main phytosociological associations present in the woods are probably Ulmo-Fraxinetum E Sjogren ap KL 1973, Querco-Fraxinetum Klötzli 1970 and Alno-Fraxinetum KP ap Seibert 1969.

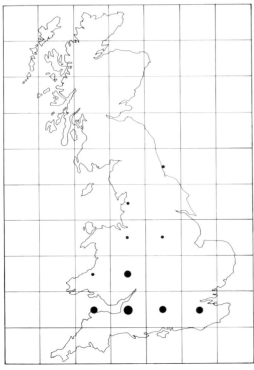

A wood on a gentle slope in southern England, the canopy in this section consists entirely of *Fagus sylvatica* (beech) but *Fraxinus excelsior* (ash) and *Acer pseudoplatanus* (sycamore) occur elsewhere in the wood.

Type 6. *Mercurialis perennis/Fraxinus-Quercus* (dog's mercury/ash-oak)

5. *Glechoma hederacea/Mercurialis perennis* (ground ivy/dog's mercury) type
A type of medium heterogeneity with *Geum urbanum* (herb bennet), *Brachypodium sylvaticum* (slender false-brome) and *Urtica dioica* (stinging nettle) as selective species. The canopy is of average density with ash and oak as the main species, and an understorey of hazel is usually present. *Mercurialis perennis* (dog's mercury) and *Rubus fruticosus* (bramble) are the usual ground cover species. The type would probably be called mixed deciduous woodland and the soils are mainly calcareous brown earths, with gleying sometimes present.

8. *Mercurialis perennis/Rubus fruticosus* (dog's mercury/bramble) type
A type of low heterogeneity with *Taxus baccata* (yew), *Carex flacca* (carnation-grass) and *Convallaria majalis* (lily-of-the-valley) as selective species. The ground cover is usually *Rubus fruticosus* (bramble) and *Mercurialis perennis* (dog's mercury). The canopy is quite dense with various combinations of ash, oak, beech and yew. A dense understorey of hazel is usually present. The type would probably be called ash woodland on limestone. The soils are mainly brown earths but there are some rendzinas.

1. *Urtica dioica/Rubus fruticosus* (stinging nettle/bramble) type
A type of low heterogeneity with *Sambucus nigra* (elder), *Mercurialis perennis* (dog's mercury) and *Euonymus europaeus* (spindle-tree) as selective species. The canopy is usually dense consisting of oak and beech with saplings often present, as well as an understorey of hazel. *Rubus fruticosus* (bramble), *Mercurialis perennis* (dog's mercury) and *Hedera helix* (ivy) are the usual ground cover species. The type would probably be referred to as mixed deciduous woodland and the soils are mainly eutrophic brown earths.

7. *Carex sylvatica/Rubus fruticosus* (wood sedge/bramble) type
A type of medium heterogeneity with *Acer campestre* (field maple), *Sorbus torminalis* (wild service tree) and *Euphorbia amygdaloides* (wood spurge) as selective species. The ground cover usually consists of *Rubus fruticosus* (bramble) and *Mercurialis perennis* (dog's mercury). The canopy is of average density consisting mainly of oak and ash with a dense understorey of hazel present in most areas. The type would probably be called mixed deciduous woodland. The soils are usually brown earths, although there is a tendency for gleying to take place.

MERCURIALIS PERENNIS/FRAXINUS-QUERCUS (DOG'S MERCURY/ASH-OAK)
TYPE

VEGETATION

Key species
Constant species: *Rubus fruticosus* (bramble), *Fraxinus excelsior* (ash), *Mercurialis perennis* (dog's mercury), *Viola riviniana* (common violet)

Plot dominants: *Rubus fruticosus* (bramble), *Mercurialis perennis* (dog's mercury), *Chamaenerion angustifolium* (rosebay willow-herb), *Hedera helix* (ivy)

Selective species: *Mercurialis perennis* (dog's mercury), *Fraxinus excelsior* (ash), *Brachypodium sylvaticum* (slender false-brome), *Viola riviniana* (common violet), *Fissidens taxifolius, Eurhynchium striatum*

Blend of   Frequency 5, 8, 1, 7, 11, 12 (4, 10)
plot types: Mean number 6.7 (high)

Mean number of species: 105 (med)
Total number of species: 222 (med)

Canopy and understorey species
Constant trees   *Constant saplings*
Ash, (sycamore, oak, birch)   Ash, sycamore (average density)

Constant shrubs   *Trees (basal area)*
(Hazel) (average density)   (Ash, oak, beech, sycamore) (average canopy)

ENVIRONMENT

Geographical distribution	Solid geology	Rainfall (cm)
SE, (NW, SW, ME, NE)	Carb li/Mg li (B, C, E)	91 *(low)*

Altitude (m)	Altitude (bot)	Altitude (top)	Slope (°)	Soil (pH)	LOI
104 *(low)*	67 m *(low)*	142 m *(low)*	14 *(low)*	6.1 *(high)*	17.7 *(low)*

SITE TYPE 7

GALIUM APARINE/ULMUS PROCERA-FRAXINUS (GOOSEGRASS/ENGLISH ELM-ASH) TYPE

A very uniform type most closely related to types 8 and 6, the former being more disturbed and less likely to be primary, whereas the latter is on less heavy soil. Plot type 3 [*Agrostis stolonifera/Mercurialis perennis* (creeping bent-grass/dog's mercury)] and type 6 [*Listera ovata/Hedera helix* (twayblade/ivy)] are locally important in most woods, with few other types present. Outside the rides and glades, this type has relatively few species present.

Ash is the main canopy species, although English elm and oak are widespread in some woods. Field maple and sycamore are also common, as well as several other minor species such as lime. Regeneration is not particularly common but hawthorn is widespread, followed closely by ash and sycamore. The canopy is often dense with few glades and natural openings, but there is vigorous vegetation by rides, or after coppicing. The variety of natural habitats is restricted as there is little variability in the ground surface. Some pheasant rearing takes place and there is often sporting interest. Grazing by domestic animals is not usual and is likely to occur only in isolated cases, because the surrounding land use is mainly arable. The woods are often isolated from each other and have other land uses nearby.

The sites usually occupy almost level land with a very narrow altitudinal range, and are usually old coppices, often with standards, now neglected. Return to high forest is therefore taking place, but Dutch elm disease is having an important impact in some woods.

This type occurs mainly in the southern England lowlands on suitable formations and is particularly common in East Anglia, becoming less common towards the north and west of its range.

Within the *Nature conservation review*, sites such as some of those in Bedford Purlieus (East Anglia), Bardney Forest (East Anglia) and Wye and Crundale (south-east England) are likely to be included in this type.

Such sites are likely to be widespread in southern Sweden, Denmark, northern Germany, Belgium and lowland France, as the vegetation has pronounced affinities with many Continental woods. However, the abundance of bramble in the ground vegetation will gradually decline away from the Atlantic influence.

The sites in this type would probably be called, in general terms, primary East Anglian boulder clay woods, or mixed deciduous woods mainly dominated by ash over dog's mercury and bramble. In comparison with most of the types, these sites are relatively uniform in phytosociological terms, probably being dominated by Ulmo-Fraxinetum E Sjogren ap KL 1973 and Querco-Fraxinetum Klötzli 1970.

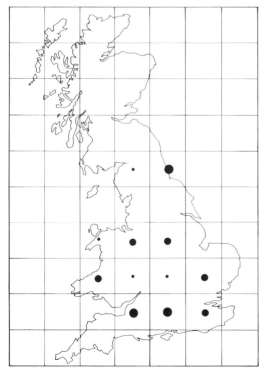

A wood in eastern England situated on almost level ground. The canopy is mainly *Quercus* spp. (oak) but *Acer pseudoplatanus* (sycamore) and *Fraxinus excelsior* (ash) are also present. Dead *Ulmus procera* (English elm) are present on the left of the picture.

Type 7.   *Galium aparine/ Ulmus procera-Fraxinus* (goosegrass/English elm-ash)

3.   *Agrostis stolonifera/Mercurialis perennis* (creeping bent-grass/dog's mercury) type
A type of very low heterogeneity with *Ulmus procera* (English elm), *Sambucus nigra* (elder) and *Acer campestre* (field maple) as selective species. The canopy is usually dense with English elm, ash and field maple as the most common species over an understorey of hazel and elder. *Mercurialis perennis* (dog's mercury) and *Agrostis stolonifera* (creeping bent-grass) are the usual ground cover species. The type would usually be termed English elm-ash woodland growing on calcareous clay. The soil is invariably a calcareous gley.

6.   *Listera ovata/Hedera helix* (twayblade/ivy) type
A type of low heterogeneity with *Ligustrum vulgare* (common privet), *Iris foetidissima* (stinging nettle) and *Viola odorata* (sweet violet) as selective species. There is usually a dense canopy with oak and ash as major species and an understorey of hazel is usually present. The type would probably be termed ash woodland mixed with pedunculate oak. The soils are predominantly calcareous brown earths.

5.   *Glechoma hederacea/Mercurialis perennis* (ground ivy/dog's mercury) type
A type of medium heterogeneity with *Geum urbanum* (herb bennet), *Brachypodium sylvaticum* (slender false-brome) and *Urtica dioica* (stinging nettle) as selective species. The canopy is of average density with ash and oak as the main species, and an understorey of hazel is usually present. *Mercurialis perennis* (dog's mercury) and *Rubus fruticosus* (bramble) are the usual ground cover species. The type would probably be called mixed deciduous woodland and the soils are mainly calcareous brown earths, with gleying sometimes present.

1.   *Urtica dioica/Rubus fruticosus* (stinging nettle/bramble) type
A type of low heterogeneity with *Sambucus nigra* (elder), *Mercurialis perennis* (dog's mercury) and *Euonymus europaeus* (spindle-tree) as selective species. The canopy is usually dense consisting of oak and beech with saplings often present, as well as an understorey of hazel. *Rubus fruticosus* (bramble), *Mercurialis perennis* (dog's mercury) and *Hedera helix* (ivy) are the usual ground cover species. The type would probably be referred to as mixed deciduous woodland and the soils are mainly eutrophic brown earths.

SUMMARY OF **SITE TYPE 7**

GALIUM APARINE/ULMUS PROCERA-FRAXINUS (GOOSEGRASS/ENGLISH ELM-ASH) TYPE

VEGETATION

Key species

Constant species: *Fraxinus excelsior* (ash), *Circaea lutetiana* (enchanter's night-shade), *Urtica dioica* (stinging nettle), *Eurhynchium praelon-gum*

Plot dominants: *Mercurialis perennis* (dog's mercury), *Hedera helix* (ivy), *Rubus fruticosus* (bramble), *Glechoma hederacea* (ground ivy)

Selective species: *Iris foetidissima* (stinking iris), *Ulmus carpinifolia* (smooth elm), *Ulmus procera* (English elm), *Ligustrum vulgare* (common privet), *Thamnium alopecurum, Fissidens taxifolius*

Blend of   Frequency 3, 6, 5 (1, 4)
plot types: Mean number 3.8 (low)

Mean number of species: 60 (low)
Total number of species: 116 (low)

Canopy and understorey species

Constant trees
Ash (English elm, oak, field maple, sycamore)

Constant saplings
Hawthorn (average density), (sycamore)

Constant shrubs
Hazel (elder) (low density)

Trees (basal area)
English elm (oak, ash, sycamore, smooth elm) (dense canopy)

ENVIRONMENT

Geographical distribution	Solid geology	Rainfall (cm)
ME (SW)	Calc clay, K marl/Lias	76 (*low*)

Altitude (m)	Altitude (bot)	Altitude (top)	Slope (°)	Soil (pH)	LOI
58 (*low*)	36 (*low*)	67 (*low*)	6.6 (*low*)	6.7 (*high*)	16.7 (*low*)

63

SITE TYPE 8

URTICA DIOICA/FRAXINUS-QUERCUS (STINGING NETTLE/ASH-OAK) TYPE

A uniform type most closely related to types 7 and 6, the former being less disturbed and more often primary whereas the latter occupies less heavy soils in general. Plot type 5 [*Glechoma hederacea/Mercurialis perennis* (ground ivy/dog's mercury)] predominates in some woods but elsewhere type 2 [*Bromus ramosus/Mercurialis perennis* (hairy brome/dog's mercury)] is common. These sites are amongst the poorest in species throughout the whole series. Ash is the major canopy species, although hawthorn, sycamore, English elm and field maple are locally abundant, as well as scattered specimens of other species. Regeneration is not usually abundant, because the canopy is often dense, but, where present, consists mainly of hawthorn and ash. The soils are either calcareous brown earths or calcareous gleys and are very eutrophic with high nitrogen and phosphorus levels and virtually no surface litter. Otherwise, brown earths are locally present.

The woods are often used as pheasant coverts, with glades maintained for management purposes. The glades and rides have rich tall-herb vegetation, often with stinging nettle and rosebay willow-herb as major species. In most sites, few additional habitats are present because of the uniform ground surface, and the woods are often small. In general, grazing by domestic stock is absent, although cattle may sometimes be present.

The woods are usually surrounded by intensely farmed arable land; otherwise, short-term grass is the most frequent land use. The sites usually occupy almost flat land or, at most, gentle slopes with a very narrow altitudinal range. Some sites are mainly old coppices but they are more frequently secondary in origin, often having a poor structure, with a sparse understorey.

This type occurs widely throughout the lowlands of England and Wales but with an emphasis on East Anglia and south-west England.

Within the *Nature conservation review*, sites such as Sapperton Pickworth (East Anglia), Ashen Copse (south-west England) and Kingley Vale (south-east England) are likely to be included in this type.

Such sites are likely to be present throughout the northern lowlands of the Continent, from Denmark through northern Germany to France, although they will have progressively more Continental species away from the Atlantic margin.

The sites in this type would probably be called, in general terms, small secondary basiphilous woods, or typically ash woods over dog's mercury and bramble. In comparison with most of the types, these sites are relatively uniform in phytosociological terms probably being dominated by Ulmo-Fraxinetum E Sjogren ap KL 1973 and Querco-Fraxinetum Klötzli 1970.

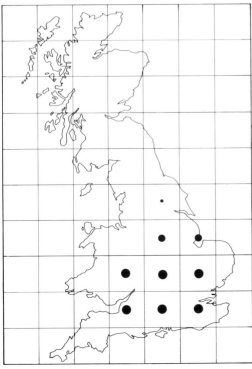

A small wood on gently sloping land in the midlands of England. The canopy is mainly *Acer pseudoplatanus* (sycamore) and *Fraxinus excelsior* (ash) but some trees of *Populus* spp. (poplar) and *Quercus* spp. (oak) are also present. A dead *Ulmus procera* (English elm) is present in the front of the wood.

Type 8. *Urtica dioica/ Fraxinus-Quercus* (stinging nettle/oak-ash)

5. *Glechoma hederacea/Mercurialis perennis* (ground ivy/dog's mercury) type
A type of medium heterogeneity with *Geum urbanum* (herb bennet), *Brachypodium sylvaticum* (slender false-brome) and *Urtica dioica* (stinging nettle) as selective species. The canopy is of average density with ash and oak as the main species, and an understorey of hazel is usually present. *Mercurialis perennis* (dog's mercury) and *Rubus fruticosus* (bramble) are the usual ground cover species. The type would probably be called mixed deciduous woodland and the soils are mainly calcareous brown earths, with gleying sometimes present.

2. *Bromus ramosus/Mercurialis perennis* (hairy brome/dog's mercury) type
A type of low heterogeneity, with *Anthriscus sylvestris* (cow parsley), *Campanula trachelium* (bats-in-the-belfrey) and *Silene dioica* (red campion) as selective species. The canopy is usually dense consisting of ash, English elm and oak, with an understorey often present. *Rubus fruticosus* (bramble), *Mercurialis perennis* (dog's mercury) and *Galeobdolon luteum* (yellow archangel) are the usual ground cover species. The type would probably be termed pedunculate oak-ash woodland growing under moist bare rich conditions. The soils are mainly brown earths with some gleys.

13. *Chrysosplenium oppositifolium/Mercurialis perennis* (opposite-leaved golden saxifrage/dog's mercury) type
A type of medium heterogeneity and an average species complement, with *Silene dioica* (red campion), *Campanula latifolia* (large campanula) and *Heracleum sphondylium* (hogweed) as selective species. The canopy is usually quite dense, with ash as the major species, although sycamore and wych elm are widespread. There is an infrequent understorey but saplings are often present in low densities. The ground cover is variable, with *Mercurialis perennis* (dog's mercury), *Rubus fruticosus* (bramble) and *Urtica dioica* (stinging nettle) being most frequent. The type is mainly ash woodland growing on steep slopes. The soils are mainly alluvial brown earths, although there is often gleying and waterlogging.

1. *Urtica dioica/Rubus fruticosus* (stinging nettle/bramble) type
A type of low heterogeneity with *Sambucus nigra* (elder), *Mercurialis perennis* (dog's mercury) and *Euonymus europaeus* (spindle-tree) as selective species. The canopy is usually dense consisting of oak and beech with saplings often present, as well as an understorey of hazel. *Rubus fruticosus* (bramble), *Mercurialis perennis* (dog's mercury) and *Hedera helix* (ivy) are the usual ground cover species. The type would probably be referred to as mixed deciduous woodland and the soils are mainly eutrophic brown earths.

URTICA DIOICA/FRAXINUS-QUERCUS (STINGING NETTLE/ASH-OAK) TYPE

VEGETATION

Key species

Constant specie:	*Urtica dioica* (stinging nettle), *Poa annua* (annual poa), *Rubus fruticosus* (bramble), *Eurhynchium praelongum*
Plot dominants:	*Mercurialis perennis* (dog's mercury), *Urtica dioica* (stinging nettle), *Rubus fruticosus* (bramble), *Galeobdolon luteum* (yellow archangel)
Selective species:	*Campanula trachelium* (bats-in-the-belfry), *Urtica dioica* (stinging nettle), *Bromus ramosus* (hairy brome), *Glechoma hederacea* (ground ivy), *Poa trivialis* (rough meadow-grass), *Brachythecium rutabulum*
Blend of plot types:	Frequency 5, 2, 13 (1, 3, 12) Mean number 4.6 (low)

Mean number of species: 79 (low)
Total number of species: 160 (low)

Canopy and understorey species

Constant trees (Ash)	*Constant saplings* Hawthorn (ash, English elm)
Constant shrubs Hazel (elder) (average density)	*Trees (basal area)* (Ash, oak) (average canopy)

ENVIRONMENT

Geographical distribution	Solid geology	Rainfall (cm)
ME, (SE, NE)	(A, B, E, G)	71 *(low)*

Altitude (m)	Altitude (bot)	Altitude (top)	Slope (°)	Soil (pH)	LOI
79 *(low)*	56 m *(low)*	96 m *(low)*	7 *(low)*	6.0 *(high)*	12.4 *(low)*

SITE TYPE 9

PTERIDIUM AQUILINUM/QUERCUS-FAGUS (BRACKEN/OAK-BEECH) TYPE

A quite variable type most closely related to site types 1 and 2, the former having a more lowland character and being more associated with clay soils, whereas the latter has an additional basiphilous element as well as being generally more upland and western. Although plot type 17 [*Pteridium aquilinum/Rubus fruticosus* (bracken/ bramble)] occurs most frequently, type 18 [*Deschampsia flexuosa/Pteridium aquilinum* (wavy hair-grass/bracken)] is also widespread. A range of other types is present, reflecting the variability of the sites.

Oak is the major canopy species, but birch and beech are locally dominant, whereas rowan and sycamore occupy lesser roles. Other species, eg aspen and willow, are present in appropriate localities. Regeneration is not usually abundant and birch is the only common species. However, oak, rowan and sycamore occur locally. The soils are mainly acidic brown earths and brown podzolics, but brown earths are also present, mainly on the lower slopes, there being marked catenas in many sites.

A wide range of habitats is present, varying from small streams to rocky banks and glades of various sizes. The latter have well-developed tall-herb vegetation often with rosebay willow-herb, bramble and bracken. The woods are often grazed by deer and some sites are open to domestic stock, both sheep and cattle, so that the grazing pressure is somewhat variable. Some woods are used for pheasant shooting.

The surrounding land use is variable but arable and permanent pasture predominate, although woodlands and housing are also frequent. The woods occupy various land forms, most usually on valley sides with a variable altitudinal range, but also on more or less level or gently sloping ground indicating convergence – where different conditions in combination lead to similar vegetation.

There is a wide range of forms of management from coppices through to colonizing woodland and to high forest. Many of the woods have groups of conifers planted but there has been relatively little recent management.

Apart from a likely high concentration in the Weald, this type has a generally western bias in its distribution being present in south-west England and Wales. It does, however, extend into north-west and north-east England, as well as having outliers elsewhere in the English lowlands and north-east Scotland.

Within the *Nature conservation review*, sites such as some woods of the Bovey Valley Woods (south-west England), Nags Head enclosure (south-west England) and Coed Gorswen (north Wales) are likely to be included in this type.

The type is likely to occur widely throughout the sub-Atlantic margins of northern Europe but progressively less frequently under more Continental conditions. The sites in this type would probably be called, in general terms, freely drained lowland acid woods or natural oak/birch woods over bracken and bramble. The main phytosociological associations present in the woods are probably Blechno-Quercetum Br-Bl et Tx 1952, Fago-Quercetum petraeae Tx 1955 and Betulo-Quercetum Tx 1937.

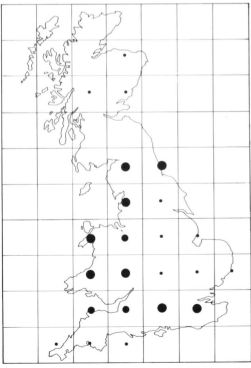

A wood on level ground in north-eastern England. *Fagus sylvatica* (beech), of planted origin, is in the foreground; otherwise, *Quercus* spp. (oak) is the main canopy species, with *Acer pseudoplatanus* (sycamore) and *Fraxinus excelsior* (ash) also present.

Type 9. *Pteridium aquilinum/Quercus-Fagus* (bracken/oak-beech)

17. *Pteridium aquilinum/Rubus fruticosus* (bracken/bramble) type
A type of low heterogeneity with *Ilex aquifolium* (holly), *Fagus sylvatica* (beech) and *Carpinus betulus* (hornbeam) as selective species. The canopy is usually dense consisting of oak and beech, but there are few saplings or shrubs. *Rubus fruticosus* (bramble), *Pteridium aquilinum* (bracken) and *Hedera helix* (ivy) are the usual ground cover species. The type would probably be termed dry, acid sessile oak woodland and the soils are mainly brown earths.

18. *Deschampsia flexuosa/Peteridium aquilinum* (wavy hair-grass/bracken) type
A type with low heterogeneity and a low species complement, *Vaccinium myrtillus* (bilberry) and *Lonicera periclymenum* (honeysuckle) being selective species. The canopy of oak and birch with some rowan is of medium density, with few saplings or shrubs. The ground cover is usually high with *Pteridium aquilinum* (bracken), *Deschampsia flexuosa* (wavy hair-grass) and *Vaccinium myrtillus* (bilberry) the most common species. The type would probably be termed acid sessile oak or birch woodland. The soils are mainly brown podzols or are podzolic in character.

21. *Oxalis acetosella/Pteridium aquilinum* (wood sorrel/bracken) type
A type with medium heterogeneity and a low species complement, *Milium effusum* (wood millet) and *Digitalis purpurea* (foxglove) being selective species. The canopy is usually dense of oak, birch or sycamore with few saplings or shrubs. The ground cover is usually *Pteridium aquilinum* (bracken), *Rubus fruticosus* (bramble) or *Holcus mollis* (creeping soft-grass). The type would probably be called acid sessile oak woodland on valley sides. The soils are usually acid brown earths.

23. *Holcus mollis/Pteridium aquilinum* (creeping soft-grass/bracken) type
A type with low heterogeneity and a low species complement, *Endymion non-scriptus* (bluebell), *Oxalis acetosella* (wood-sorrel) and *Stellaria holostea* (greater stitchwort) being selective species. The canopy, of oak and sometimes birch, is invariably dense with few saplings and some shrubs. The ground cover is mainly *Pteridium aquilinum* (bracken), *Rubus fruticosus* (bramble), and *Dryopteris dilatata* (broad buckler-fern). The type would probably be termed acidic oak/birch woodland on rather heavy soils. The soils are mainly acid brown earths, often with some evidence of gleying.

SUMMARY OF **SITE TYPE 9**
PTERIDIUM AQUILINUM/QUERCUS-FAGUS (BRACKEN/OAK-BEECH) TYPE

VEGETATION

Key species
Constant species: *Rubus fruticosus* (bramble), *Quercus* spp. (oak), *Pteridium aquilinum* (bracken), *Lonicera periclymenum* (honeysuckle)

Plot dominants: *Pteridium aquilinum* (bracken), *Rubus fruticosus* (bramble), *Holcus mollis* (creeping soft-grass), *Deschampsia flexuosa* (wavy hair-grass)

Selective species: *Chamaenerion angustifolium* (rosebay willow-herb), *Fagus sylvatica* (beech), *Ilex aquifolium* (holly), *Deschampsia flexuosa* (wavy hair-grass), *Pteridium aquilinum* (bracken), *Dicranella heteromalla*

Blend of   Frequency 17, 18, 21, 23 (9, 10, 12, 24)
plot types: Mean number 6.1 (med)

Mean number of species: 88 (low)
Total number of species: 288 (high)

Canopy and understorey species
Constant trees   *Constant saplings*
Oak (birch)   Birch (low density)

Constant shrubs   *Trees (basal area)*
—   Oak (beech, birch) (average canopy)

ENVIRONMENT

Geographical distribution	*Solid geology*	*Rainfall (cm)*
NE, SE, SW (ME, NW)	Wealden, Mill grit/ Coal mea (D, B)	86 *(low)*

Altitude (m)	*Altitude (bot)*	*Altitude (top)*	*Slope (°)*	*Soil (pH)*	*LOI*
112 *(low)*	68 m *(low)*	155 m *(low)*	17 *(med)*	4.2 *(low)*	14.8 *(low)*

SITE TYPE 10

TEUCRIUM SCORODONIA/QUERCUS-BETULA (WOOD SAGE/OAK-BIRCH) TYPE

This type is quite variable and is most closely related to types 9 and 14, the former being more lowland in character whereas the latter is more upland and occurs generally on steeper, rockier slopes. Plot type 25 [*Galium saxatile/Deschampsia flexuosa* (heath bedstraw/wavy hair-grass)] is consistently dominant through the woods but 18 [*Deschampsia flexuosa/Pteridium aquilinum* (wavy hair-grass/bracken)] is also widespread. Other types with more basiphilous affinities are associated with streamsides and flush lines.

Oak is the major canopy species, although birch is also abundant, particularly on disturbed sites. Otherwise, sycamore, ash and rowan are locally dominant in patches in the woods. Regeneration is not usually abundant, because of grazing pressure, but, where present, is mainly of birch, although rowan and oak were also commonly recorded.

The soils are mainly acid brown earths, often very stony and rocky. However, brown podzolics and even some podzols are present, particularly towards the upper level of the woods. A wide range of habitats is present – particularly those associated with rock, eg mossy boulders and outcrops, as well as a range of aquatic features. Glades and clearings are scattered throughout the woods, usually grassy or covered with bracken. The woods are generally grazed by various combinations of deer, sheep and cattle, according to the state of the fencing which is highly variable.

The land use below the woods is usually permanent pasture, with other woodlands also common nearby. Above the woods, rough pasture or moorland is the usual land use. The woods occupy steep hillsides with complex topography and a wide altitudinal range, and are set in upland landscapes. Most of the sites are old coppice woods, used for charcoal and tan bark. Some have been singled and are returning to high forest, but elsewhere the stems have developed from neglected coppice stools.

The centres of distribution of this type are west Wales, and the Lake District. However, the type is widespread in western Britain, especially on Dartmoor and Exmoor, and extends into the west of Scotland, particularly Argyllshire. Outliers are probably present elsewhere in Scotland and also in the Weald.

Within the *Nature conservation review*, sites such as Naddle Forest (north-west England), Holne Chase (south-west England) and Coed Rheidol (south Wales) are likely to be included in this type.

On the Continent, it is limited to the more upland areas of the Atlantic margin, particularly western Norway, Brittany and the foothills of the Pyrenees, and perhaps the Massif Central. The type is also probably represented in Ireland, in areas such as the Wicklow mountains.

The sites in this type would probably be called, in general terms, western acidic hillsides, or oak-birch woods over bracken and wavy hair-grass. The main phytosociological associations present in the woods are probably Blechno-Quercetum Br-Bl et Tx 1952, Galio-saxatilis-Quercetum Birse et Robertson 1976 and Dryopterido-Fraxinetum Klötzli 1970.

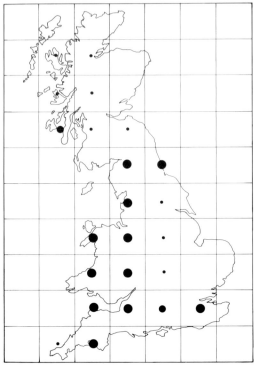

A wood occupying a valley side in the English Lake District. The canopy composition is mainly *Quercus* spp. (oak), but *Betula* spp. (birch) is widespread and both *Fraxinus excelsior* (ash) and *Alnus glutinosa* (alder) grow by streamsides in the wood.

Type 10. *Teucrium scorodonia/Quercus-Betula* (wood sage/oak-birch)

25. *Galium saxatile/Deschampsia flexuosa* (heath bedstraw/wavy hair-grass) type
A type with medium heterogeneity and a low species complement with *Agrostis canina* (brown bent-grass), *Anthoxanthum odoratum* (sweet vernal-grass) and *Vaccinium myrtillus* (bilberry) as selective species. The canopy is usually of average density, oak being the major species but birch is also common. Few saplings are present, and an understorey is likewise scarce. The ground cover is usually of *Deschampsia flexuosa* (wavy hair-grass), *Pteridium aquilinum* (bracken) or *Agrostis tenuis* (common bent-grass). The type would be called western acid sessile oak wood. The soils are mainly brown podzolic in character.

18. *Deschampsia flexuosa/Pteridium aquilinum* (wavy hair-grass/bracken) type
A type with low heterogeneity and a low species complement, *Vaccinium myrtillus* (bilberry) and *Lonicera periclymenum* (honeysuckle) being selective species. The canopy of oak and birch with some rowan is of medium density, with few saplings or shrubs. The ground cover is usually high, with *Pteridium aquilinum* (bracken), *Deschampsia flexuosa* (wavy hair-grass) and *Vaccinium myrtillus* (bilberry) the most common species. The type would probably be termed acid sessile oak or birch woodland. The soils are mainly brown podzols or are podzolic in character.

10. *Athyrium filix-femina/Rubus fruticosus* (lady-fern/bramble) type
A type of medium heterogeneity with *Circaea lutetiana* (enchanter's nightshade), *Veronica montana* (wood speedwell) and *Dryopteris dilatata* (broad buckler-fern) as selective species. The ground cover usually consists mainly of *Rubus fruticosus* (bramble). The canopy is dense, mainly of oak and ash but also sycamore, birch, willow and alder with few saplings, and an understorey of hazel is often present. The type would probably be termed moist pedunculate oak-ash woodland. The soils are mainly brown earths often rather rocky and sometimes shallow.

21. *Oxalis acetosella/Pteridium aquilinum* (wood-sorrel/bracken) type
A type with medium heterogeneity and a low species complement, *Milium effusum* (wood millet) and *Digitalis purpurea* (foxglove) being selective species. The canopy is usually dense of oak, birch or sycamore with few saplings or shrubs. The ground cover is usually *Pteridium aquilinum* (bracken), *Rubus fruticosus* (bramble) or *Holcus mollis* (creeping soft-grass). The type would probably be called acid sessile oak woodland on valley sides. The soils are usually acid brown earths.

TEUCRIUM SCORODONIA/QUERCUS-BETULA (WOOD SAGE/OAK-BIRCH) TYPE

VEGETATION

Key species
Constant species: *Oxalis acetosella* (wood-sorrel), *Quercus* spp. (oak), *Dryopteris dilatata* (broad buckler-fern), *Rubus fruticosus* (bramble)

Plot dominants: *Deschampsia flexuosa* (wavy hair-grass), *Pteridium aquilinum* (bracken), *Rubus fruticosus* (bramble), *Agrostis tenuis* (common bent-grass)

Selective species: *Oxalis acetosella* (wood-sorrel), *Luzula pilosa* (hairy wood-rush), *Teucrium scorodonia* (wood sage), *Polytrichum* spp., *Plagiothecium undulatum*, *Dicranum scoparium*

Blend of Frequency 25, 18, 10, 21, 22, 23 (9, 11, 26, 29)
plot types: Mean number 5.9 (med)

Mean number of species: 94 (med)
Total number of species: 195 (med)

Canopy and understorey species

Constant trees	*Constant saplings*
Oak, birch	Birch (average density)

Constant shrubs	*Trees (basal area)*
(Hazel) (low density)	Oak (birch) (average canopy)

ENVIRONMENT

Geographical distribution	Solid geology	Rainfall (cm)
NW (Wa, SW)	Silur/Ordov	145 *(high)*

Altitude (m)	Altitude (bot)	Altitude (top)	Slope (°)	Soil (pH)	LOI
131 *(med)*	85 m *(med)*	207 m *(med)*	30.0 *(high)*	4.2 *(low)*	23.4 *(med)*

SITE TYPE 11

ATHYRIUM FILIX-FEMINA/QUERCUS-FRAXINUS (LADY-FERN/OAK-ASH) TYPE

This type is the most complex of the series and is most closely related to site types 12 and 16, the former being rather more upland and dominated by vegetation associated with high water levels, whereas the latter has more northern affinities, but both are less diverse. The most extensive plot types are 22 [*Blechnum spicant/Rubus fruticosus* (hard fern/bramble)], type 12 [*Geum urbanum/Mercurialis perennis* (herb bennet/dog's mercury)] and type 21 [*Oxalis acetosella/Pteridium aquilinum* (wood-sorrel/bracken)]. There is, however, a wide range of other types scattered through the woods, depending upon local topographic variation such as seepage lines and rock outcrops.

Oak and ash are the main canopy species but birch and rowan occur in the drier areas whereas alder and willow are associated with the wetter conditions, as well as other species such as aspen. The soils are likewise variable, with acid brown earths and brown earths predominating in well-drained areas, but a range of gleys and gleyed brown earths is present under wetter conditions. The habitats are very variable reflecting the complex topography and include features such as riversides, cliffs, rock ledges and marshland. Although often grazed by sheep, cattle and deer, the steep gorge sides are usually inaccessible, and thus present strong contrasts in grazing pressure within the woodland boundary. In the ungrazed areas, there is often abundant regeneration consisting usually of ash, rowan and oak.

Although the most common surrounding land use is pasture, there is often arable nearby as well as moorland vegetation in the upper sections of more upland sites. The majority of sites occupy precipitous gorges or at least steep riversides with a very wide altitudinal range. They are usually in marginal situations between uplands and lowlands.

Because of the difficulty of access, the woods have generally been exploited only in a haphazard fashion for firewood, or timber, in more accessible locations. Their high landscape value has meant that there has often been some amenity planting, and groups of conifers and other exotics are often scattered through the sites.

Although most of the sites are likely to be in western Britain, there are outliers elsewhere in Scotland and north-east England, which are likely to be repeated on topographically suitable sites. Nevertheless, the type is likely to be most frequent in the Lake District and west Wales.

Within the *Nature conservation review*, sites such as Roeburndale (north-west England), Keltney Burn (east Scotland) and Ceunant Llanerych (north Wales) are likely to be included in this type.

On the Continent, such sites are likely to be restricted to western Norway and the foothills of the Pyrenees, although some other sites could be present in isolated areas elsewhere, eg the Massif Central, under suitable topographic conditions.

The sites of this type would probably be called, in general terms, ash-oak woods in complex gorges with heterogeneous vegetation. The main phytosociological associations present in the woods are probably Carici-remotae-Fraxinetum (W Koch 1926) Schwickerath 1937 and Br-Bl et Tx 1952, but several other types are also likely to be present.

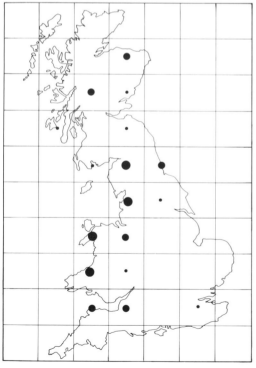

A wood occupying the steep sides of a gorge in central Scotland. There is a plantation of *Fagus sylvatica* (beech) outside the gorge, but, within, *Fraxinus excelsior* (ash), *Betula* spp. (birch), *Quercus* spp. (oak) and *Corylus avellana* (hazel) predominate.

Type 11. *Athyrium filix-femina/Quercus-Fraxinus* (lady-fern/oak-ash)

12.   *Geum urbanum/Mercurialis perennis* (herb bennet/dog's mercury) type
A type of medium heterogeneity and a high species complement, with *Brachypodium sylvaticum* (slender false-brome), *Fragaria vesca* (wild strawberry) and *Sanicula europea* (sanicle) as selective species. The canopy is usually quite dense with oak and ash as major species, but birch, sycamore, alder and hawthorn are also present locally. There is an understorey of hazel, with saplings of low density but frequent occurrence. The ground cover is principally *Mercurialis perennis* (dog's mercury) and *Rubus fruticosus* (bramble). The type belongs to the broad range of pedunculate oak-ash woodland on quite basiphilous valley sides. The soils are mainly brown earths, often skeletal, and usually gleyed.

22.   *Blechnum spicant/Rubus fruticosus* (hard fern/bramble) type
A type with medium heterogeneity and an average species complement, *Luzula sylvatica* (greater woodrush), *Athyrium filix-femina* (lady-fern) and *Hedera helix* (ivy) being selective species. The canopy is of average density, mainly of oak but with birch, rowan, beech, ash and sycamore also present, and a few saplings and some shrubs. The ground cover is mainly *Rubus fruticosus* (bramble) and *Luzula sylvatica* (greater woodrush), but there is often a dense carpet of leaves. The type would usually be called mixed deciduous woodland on steep valley sides. The soils are mainly acid brown earths and often very rocky.

10.   *Athyrium filix-femina/Rubus fruticosus* (lady fern/bramble) type
A type of medium heterogeneity with *Circaea lutetiana* (enchanter's nightshade), *Veronica montana* (wood speedwell) and *Dryopteris dilatata* (broad buckler-fern) as selective species. The ground cover usually consists mainly of *Rubus fruticosus* (bramble). The canopy is dense, mainly of oak and ash but also sycamore, birch, willow and alder with few saplings, and an understorey of hazel is often present. The type would probably be termed moist pedunculate oak-ash woodland. The soils are mainly brown earths, although often rather rocky and sometimes shallow.

26.   *Potentilla erecta/Holcus mollis* (common tormentil/creeping soft-grass) type
A type of medium heterogeneity and an average species complement, with *Galium saxatile* (heath bedstraw), *Anthoxanthum odoratum* (sweet vernal-grass) and *Deschampsia flexuosa* (wavy hair-grass) as selective species. The canopy is usually of average density, consisting mainly of oak but with birch and rowan often present, as well as saplings and an understorey of hazel. The ground cover is usually of *Holcus mollis* (creeping soft-grass), *Pteridium aquilinum* (bracken) or *Agrostis tenuis* (common bent-grass). The type would usually be termed western acid sessile oak woodland but has some enrichment. The soils vary in character from acid brown earths to brown podzolics.

VEGETATION

Key species
Constant species: *Oxalis acetosella* (wood-sorrel), *Fraxinus excelsior* (ash), *Quercus* spp. (oak), *Dryopteris filix-mas* (male fern), *Luzula sylvatica* (greater woodrush), *Pteridium aquilinum* (bracken)

Plot dominants: *Holcus mollis* (creeping soft-grass), *Dryopteris filix-mas* (male fern)

Selective species: *Luzula sylvatica* (greater woodrush), *Rubus idaeus* (raspberry), *Athyrium filix-femina* (lady-fern), *Oxalis acetosella* (wood-sorrel), *Digitalis purpurea* (foxglove), *Sorbus aucuparia* (rowan)

Blend of   Frequency 12, 22, 10, 26 (11, 13, 14, 16, 21, 25, 29, 30)
plot types: Mean number 8.1 (high)

Mean number of species: 136 (high)
Total number of species: 264 (high)

Canopy and understorey species
Constant trees   *Constant saplings*
Oak (ash, birch, rowan)   —

Constant shrubs   *Trees (basal area)*
(Hazel) (low density)   Oak (ash, sycamore, birch) (average canopy)

ENVIRONMENT

Geographical distribution	*Solid geology*	*Rainfall (cm)*
ES (WS, NW, Wa, NE)	Silur/Ordov, Red s st (G)	117 *(med)*

Altitude (m)	*Altitude (bot)*	*Altitude (top)*	*Slope (°)*	*Soil (pH)*	*LOI*
124 *(med)*	113 m *(high)*	194 m *(low)*	24.0 *(med)*	4.7 *(low)*	15.9 *(low)*

SITE TYPE 12

RANUNCULUS REPENS/QUERCUS-ALNUS (CREEPING BUTTERCUP/OAK-ALDER) TYPE

A quite variable type most closely related to site types 3 and 11, the former being more lowland in its general affinities and being also on more level ground, whereas the latter is more northern and upland, as well as being more variable in ground conditions and vegetation. No one plot type predominates, although type 29 [*Anthoxanthum odoratum/Agrostis tenuis* (sweet vernal-grass/common bent-grass)] is the most common; otherwise, types 25 [*Galium saxatile/Deschampsia flexuosa* (heath bedstraw/wavy hair-grass)] and 16 [*Cirsium palustre/Agrostis tenuis* (marsh thistle/common bent-grass)] as well as 11, 12 and 14 all occur widely. There is a very strong contrast between the soil types present: those on the drier banks are mainly acid brown earths or brown podzolics, whereas in the wet areas within the woods gleys and soils with a humus-rich horizon associated with a permanent surface water cover are found.

There is a comparable contrast in the major canopy species, with oak, rowan and birch associated with the better drained areas, whereas the wet areas have alder, ash and various willow species. In general, the woods are heavily grazed by domestic stock as well as deer – although the animals avoid the very wet areas. Regeneration is thus generally scarce, although in some areas there are patches of alder and willow colonizing open areas of wet land within the boundaries of the woods.

A wide range of habitats is present, particularly those associated with the small streams or rivers that pass through the woods. Rock outcrops, boulders, and other upland features are present in the sites at higher altitudes.

Although the sites are usually on hillsides with extensive flushing, convergent conditions also occur in lowland situations that act as water-collecting areas. There is thus a strong element of variability in the land forms occupied by the sites, which is also likely to be found in further sites identified by the Key. Sometimes the sites have a wide altitudinal range but occasionally, under particular conditions, are within a limited band. The dominant feature is the extensive nature of the marshlands. Extensive open areas are thus often present. In terms of adjacent land use, woodland is common; otherwise, permanent pasture, upland grazings or moorland at the upper margins of the woods are the usual associated land uses.

The drier areas of the woods have usually been managed as coppices, but elsewhere the management has not been intensive.

Although the sites recorded to date were from the extreme west of Britain, it is possible that comparable sites will occur by estuaries elsewhere where marshland predominates. However, most of the sites will be in the uplands and are likely to be uncommon.

Within the *Nature conservation review*, sites such as Ariundle (west Scotland), Gartfairn (within the Loch Lomond woods in west Scotland) and Coedydd Aber (north Wales) are likely to be included in this type.

Elsewhere in Europe, such sites are likely to be confined to the extreme west, ie western Norway and Brittany, although comparable, very marshy woods may be found in the valley floors of the big rivers such as the Loire and the Rhine. They are also likely to be present in the west of Ireland.

The sites in this type would probably be called, in general terms, marshy woods with alder and oak over variable vegetation. Although Blechno-Quercetum Br-Bl et Tx 1952 is probably the most common background vegetation in phytosociological terms, several other types linked to the marshy conditions are likely to be present.

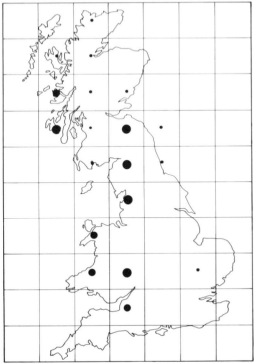

A wood in north Wales, of mixed species of *Quercus* spp. (oak), *Fraxinus excelsior* (ash), *Alnus glutinosa* (alder), with some scattered *Corylus avellana* (hazel) and *Crataegus monogyna* (hawthorn).

Type 12. *Ranunculus repens/Quercus-Alnus* (creeping buttercup/oak-alder)

29. *Anthoxanthum odoratum/Agrostis tenuis* (sweet vernal-grass/common bent-grass) type

A type with medium heterogeneity and an average species complement, *Rumex acetosa* (sorrel), *Potentilla erecta* (common tormentil) and *Galium saxatile* (heath bedstraw) being the selective species. The canopy is usually of average density, of oak or birch, with saplings and an understorey often present. The ground cover is usually *Agrostis tenuis* (common bent-grass), *Pteridium aquilinum* (bracken) or *Holcus mollis* (creeping soft-grass). The type would probably be called upland oak or birch woodland. The soils are variable, but are mainly acid brown earth or brown podzolic in character.

16. *Cirsium palustre/Agrostis tenuis* (marsh thistle/common bent-grass) type

A type of high heterogeneity and a high species complement, *Galium mollugo* (hedge bedstraw), *Cerastium holosteoides* (common mouse-ear chickweed) and *Dactylis glomerata* (cook's-foot) being selective species. The canopy is open, with few saplings or shrubs. The ground cover is mainly *Agrostis tenuis* (common bent-grass) but other species, eg *Athyrium filix-femina* (lady-fern), are locally important. The type would probably be called riverine upland ash woodland. The soils are usually gleys flushed by lateral water movement and are often shallow.

25. *Galium saxatile/Deschampsia flexuosa* (heath bedstraw/wavy hair-grass) type

A type of medium heterogeneity and a low species complement, with *Agrostis canina* (brown bent-grass), *Anthoxanthum odoratum* (sweet vernal-grass) and *Vaccinium myrtillus* (bilberry) as selective species. The canopy is usually of average density, oak being the major species, but birch is also common. Few saplings are present, and an understorey is likewise scarce. The ground cover is usually of *Deschampsia flexuosa* (wavy hair-grass), *Pteridium aquilinum* (bracken) or *Agrostis tenuis* (common bent-grass). The type would be called western acid sessile oak wood. The soils are mainly brown podzolic in character.

14. *Chrysosplenium oppositifolium/Rubus fruticosus* (opposite-leaved golden saxifrage/bramble) type

A type of medium heterogeneity and an average species complement, with *Iris pseudacorus* (yellow flag), *Galium palustre* (marsh bedstraw) and *Solanum dulcamara* (bittersweet) as selective species. The canopy is usually dense with ash and alder as the major species, although willow and birch are widespread. An understorey is usually present, with many saplings. The ground cover is principally *Rubus fruticosus* (bramble), but *Chyrsosplenium oppositifolium* (opposite-leaved golden saxifrage) and *Urtica dioica* (stinging nettle) are common also. The type is mainly ash-alder woodland growing on eutrophic valley floors. The soils are invariably heavily gleyed, often with a surface humus-rich horizon that is waterlogged.

RANUNCULUS REPENS/QUERCUS-ALNUS (CREEPING BUTTERCUP/OAK-ALDER)
TYPE

VEGETATION

Key species

Constant species: *Agrostis tenuis* (common bent-grass), *Oxalis acetosella* (wood-sorrel), *Athyrium filix-femina* (lady-fern), *Ranunculus repens* (creeping buttercup)

Plot dominants: *Agrostis tenuis* (common bent-grass), *Pteridium aquilinum* (bracken), *Holcus mollis* (creeping soft-grass), *Agrostis stolonifera* (creeping bent-grass)

Selective species: *Mentha aquatica* (water mint), *Rumex acetosa* (sorrel), *Viola palustris* (marsh violet), *Ranunculus repens* (creeping buttercup), *Anthoxanthum odoratum* (sweet vernal-grass), *Juncus effusus* (soft rush)

Blend of plot types: Frequency 29, 16, 25, 14 (11, 12, 15, 21, 22, 26, 30, 31, 32) Mean number 6.3 (med)

Mean number of species: 150 (high)
Total number of species: 240 (high)

Canopy and understorey species

Constant trees
Oak (birch, alder, ash)

Constant saplings
—

Constant shrubs
(Hazel) (low density)

Trees (basal area)
Oak (alder, birch) (average canopy)

ENVIRONMENT

Geographical distribution	*Solid geology*	*Rainfall (cm)*
Wa (WS)	Silur/Ordov (G I)	150 *(high)*

Altitude (m)	*Altitude (bot)*	*Altitude (top)*	*Slope (°)*	*Soil (pH)*	*LOI*
101 *(low)*	82 m *(med)*	349 m *(high)*	36.0 *(high)*	4.8 *(low)*	22.1 *(med)*

SITE TYPE 13

POTENTILLA ERECTA/QUERCUS-PINUS-BETULA (COMMON TORMENTIL/OAK-PINE-BIRCH) TYPE

A uniform type that is quite distinct from others in the series, although closest to types 16 and 15, both being markedly upland but the former being more variable overall and the latter having localized areas of highly eutrophic soils. Plot type 27 [*Calluna vulgaris/Pteridium aquilinum* (ling/bracken)] and type 28 [*Narthecium ossifragum/Molinia caerulea* (bog asphodel/purple moor-grass)] predominate throughout these sites in variable proportions according to local conditions. Otherwise, only limited areas of other types are present. A wide range of non-woodland species, particularly from moorlands and upland grasslands, is also present.

Because of their restricted representation in British woodlands as a whole, the sites were included within one type, although a wide range of site conditions is present and further analysis would be required to separate them further.

Birch, together with various proportions of oak or pine, dominates the canopy in most of the sites, and few other species except rowan, and occasionally aspen, are present. The soils are variable but are all pronouncedly upland in character, varying from peats to peaty gleys, through to brown podzolics into podzols, dependent largely upon altitude and the nutrient status of the substrate.

There is a wide range of habitats present, particularly asociated with the frequent small streams that pass through the woods, but many rocky areas are also present, as well as glades and open areas. The predominantly western distribution leads to a rich moss, liverwort and lichen flora, on both trees and boulders.

The sizes are invariably set in a mixture of moorland and rough grazing and are grazed by sheep, cattle and deer which usually have free access to the woods. These areas in particular are usually heavily grazed, being important for shelter. However, the lower levels on some sites have permanent pasture and also other woodlands as adjacent land uses. The intensive grazing pressure means that regeneration is usually scarce – although some permanently fenced sites show abundant regeneration, particularly of birch and rowan.

In topographic terms, the woods are usually growing on steep hillsides but also occupy a variety of situations from morraines to precipitous cliffs and streamsides in the wide range of altitudes. The patterns of previous use are variable depending on the situation, but have mainly been exploited for timber coppice and firewood.

Although primarily distributed in the uplands of Scotland, the Lake District and Wales, there are likely to be outliers on very poor lowland substrata, such as the Bagshot beds and the New Forest.

Within the *Nature conservation review*, a considerable number of sites would be included in this type ranging from the native pine wood series to Keskadale (north-west England), Nant Irfon (south Wales) and Dinnet Moor (east Scotland).

Comparable sites will probably be present on very poor soils in western Norway, western Sweden, Denmark, northern Germany through into lowland Belgium and France. However, in Britain, the moorland vegetation will be dominated by Atlantic species that will be lost in more Continental conditions. Some small sites may also be present in the extreme west of Ireland.

The sites in this type would probably be called, in general terms, upland oak, birch or pine woods over heather, bilberry and moorland ground vegetation, with acid and peaty soils. The main phytosociological associations are probably Blechno-Quercetum Br-Bl et Tx 1952 and Betulo-Quercetum Tx 1937.

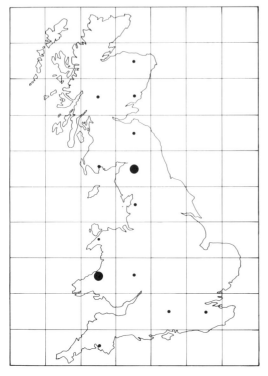

A wood on a steep slope in western Scotland principally of *Betula* spp. (birch) but also with *Quercus* (oak) scattered through the canopy.

Type 13. *Potentilla erecta/ Quercus-Pinus-Betula* (common tormentil/oak-pine-birch)

85

27. *Calluna vulgaris/Pteridium aquilinum* (ling/bracken) type
A type of medium heterogeneity and an average species complement, with *Erica cinerea* (bell-heather), *Luzula multiflora* (many-headed woodrush) and *Vaccinium myrtillus* (bilberry) as selective species. The canopy is rather open, of oak or birch, and there is rarely an understorey or saplings present. The ground cover is usually *Pteridium aquilinum* (bracken), *Deschampsia flexuosa* (wavy hair-grass) or *Calluna vulgaris* (ling). The type covers a range of traditional descriptions but is mainly birch or oak woodland in poor, freely drained conditions. The soils are mainly brown podzolics or podzols.

28. *Narthecium ossifragum/Molinia caerulea* (bog asphodel/purple moor-grass) type
A type with medium heterogeneity and an average species complement, in which *Carex echinata* (star sedge), *Drosera rotundifolia* (sundew) and *Eriophorum angustifolium* (common cotton-grass) are selective species. The canopy is usually very open, of Scots pine or birch, with few saplings or shrubs commonly present. The ground cover is usually *Molinia caerulea* (purple moor-grass), *Pteridium aquilinum* (bracken) or *Calluna vulgaris* (ling). The type contains most vegetation of native pine woods and upland birch woodland. The soils are mainly peaty podzols, peaty gleys or peats.

25. *Galium saxatile/Deschampsia flexuosa* (heath bedstraw/wavy hair-grass) type
A type of medium heterogeneity and a low species conplement, with *Agrostis canina* (brown bent-grass), *Anthoxanthum odoratum* (sweet vernal-grass) and *Vaccinium myrtillus* (bilberry) as selective species. The canopy is usually of average density, oak being the major species but birch is also common. Few saplings are present, and an understorey is likewise scarce. The ground cover is usually of *Deschampsia flexuosa* (wavy hair-grass), *Pteridium aquilinum* (bracken) or *Agrostis tenuis* (common bent-grass). The tyye would be called western acid sessile oak wood. The soils are mainly brown podzolic in character.

26. *Potentilla erecta/Holcus mollis* (common tormentil/creeping soft-grass) type
A type of medium heterogeneity and an average species complement, with *Galium saxatile* (heath bedstraw), *Anthoxanthum odoratum* (sweet vernal-grass) and *Deschampsia flexuosa* (wavy hair-grass) as selective species. The canopy is usually of average density, consisting mainly of oak but with birch and rowan often present, as well as saplings and an understorey of hazel. The ground cover is usually of *Holcus mollis* (creeping soft-grass), *Pteridium aquilinum* (bracken) or *Agrostis tenuis* (common bent-grass). The type would usually be termed western acid sessile oak woodland but has some enrichment. The soils vary in character from acid brown earths to brown podzolic series.

POTENTILLA ERECTA/QUERCUS-PINUS-BETULA (COMMON TORMENTIL/OAK-
PINE-BIRCH) TYPE

VEGETATION

Key species

Constant species: *Potentilla erecta* (common tormentil), *Pteridium aquilinum* (bracken), *Deschampsia flexuosa* (wavy hair-grass), *Galium saxatile* (heath bedstraw)

Plot dominants: *Pteridium aquilinum* (bracken), *Vaccinium myrtillus* (bilberry), *Calluna vulgaris* (ling), *Deschampsia flexuosa* (wavy hair-grass)

Selective species: *Potentilla erecta* (common tormentil), *Pinus sylvestris* (Scots pine), *Calluna vulgaris* (ling), *Erica cinerea* (bell-heather), *Molinia caerulea* (purple moor-grass), *Pleurozium schreberi*

Blend of   Frequency 27, 28 (25, 26)
plot types: Mean number 3.3 (low)

Mean number of species: 106 (med)
Total number of species: 145 (low)

Canopy and understorey species

Constant trees	*Constant saplings*
(Birch, pine, oak)	Birch (pine) (low density)

Constant shrubs	*Trees (basal area)*
—	Oak (pine) (open canopy)

ENVIRONMENT

Geographical distribution	*Solid geology*	*Rainfall (cm)*
WS (WS, Wa)	Ign/Metam, Silur/Ordov	144 *(high)*

Altitude (m)	*Altitude (bot)*	*Altitude (top)*	*Slope (°)*	*Soil (pH)*	*LOI*
195 *(high)*	132 m *(high)*	404 m *(high)*	29.0 *(high)*	4.5 *(low)*	27.1 *(high)*

SITE TYPE 14

ANTHOXANTHUM ODORATUM/QUERCUS (SWEET VERNAL-GRASS/OAK) TYPE

A quite variable type most closely related to types 13 and 10, the former being more strongly acidic and upland in nature whereas the latter has a greater predominance of brown earth soils and is less extreme. Plot types 25 [*Galium saxatile/Deschampsia flexuosa* (heath bedstraw/wavy hair-grass)] and 26 [*Potentilla erecta/Holcus mollis* (common tormentil/creeping soft-grass)] predominate throughout the majority of woods, but types 27 and 29 also occur widely in many sites.

The soils are mainly brown podzolic in character but acid brown earths are also widespread with podzols important at high levels. Oak is generally the main canopy species, but birch is also often dominant with juniper in clearings. Ash and rowan are widespread with other species such as sycamore locally important. A wide variety of upland habitats is present, such as cliffs and marshes, with streams often passing through the woods. There is a rich flora of mosses, liverworts and lichens on both boulders and trees. Open areas and glades are also widespread, with non-woodland vegetation often predominating.

The woods are mainly bounded by permanent pasture and rough grazing but moorland and other areas of woodland are also often present nearby. The sites are usually open to grazing, sheep, cattle and deer being widely present throughout and having a major effect on the ground vegetation. Regeneration is thus mainly restricted to steeper banks, and is mainly of birch and rowan but occasionally oak in more open areas. The woods are often important as shelter for animals.

The woods are invariably set in upland landscapes, mainly surrounded by steep rocky slopes, and occupy a wide altitudinal range. They are therefore often important in landscape terms. In the southern representatives, the main form of management was as coppice for charcoal but here, and elsewhere, the woods have been used for firewood and timber, although rarely intensively.

Although centred on the Lake District and Wales, this type is likely to occur widely in Scotland, particularly in the west. Although Exmoor and Dartmoor are generally not sufficiently upland in nature, some more extreme sites may fall into this type.

Within the *Nature conservation review*, sites such as some of the Borrowdale woods (north-west England), the Wood of Cree (south Scotland) and Letterewe oak woods (west Scotland) are likely to be included in this type.

On the Continent, such sites are only likely to be found on the hard old rocks of western Norway and Brittany or in the foothills of the western Pyrenees.

The sites in this type would probably be called, in general terms, mixed upland oak-birch woods on steep rocky slopes with generally poor soils. The main phytosociological associations are probably Blechno-Quercetum Br-Bl et Tx 1952, Galio-saxatilis-Quercetum Birse et Robertson 1976 and Betulo-Quercetum Tx 1937.

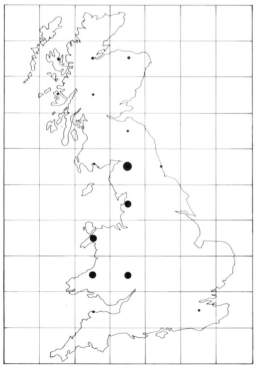

A wood in central Wales, primarily of *Quercus* spp. (oak) but with some *Betula* (birch) and a few emergent *Fraxinus excelsior* (ash) on the lower section of the slope.

Type 14. *Anthoxanthum odoratum/Quercus* (sweet vernal-grass/oak)

25. *Galium saxatile/Deschampsia flexuosa* (heath bedstraw/wavy hair-grass) type
A type of medium heterogeneity and a low species complement, with *Agrostis canina* (brown bent-grass), *Anthoxanthum odoratum* (sweet vernal-grass) and *Vaccinium myrtillus* (bilberry) as selective species. The canopy is usually of average density, oak being the major species, but birch is also common. Few saplings are present and an understorey is likewise scarce. The ground cover is usually of *Deschampsia flexuosa* (wavy hair-grass), *Pteridium aquilinum* (bracken) or *Agrostis tenuis* (common bent-grass). The type would be called western acid sessile oak wood. The soils are mainly brown podzolic in character.

26. *Potentilla erecta/Holcus mollis* (common tormentil/creeping soft-grass) type
A type of medium heterogeneity and an average species complement, with *Galium saxatile* (heath bedstraw), *Anthoxanthum odoratum* (sweet vernal-grass) and *Deschampsia flexuosa* (wavy hair-grass) as selective species. The canopy is usually of average density, consisting mainly of oak but with birch and rowan often present, as well as saplings and an understorey of hazel. The ground cover is usually of *Holcus mollis* (creeping soft-grass), *Pteridium aquilinum* (bracken) or *Agrostis tenuis* (common bent-grass). The type would usually be termed western acid sessile oak woodland but has some enrichment. The soils vary in character from acid brown earths to brown podzolics.

27. *Calluna vulgaris/Pteridium aquilinum* (ling/bracken) type
A type of medium heterogeneity and an average species complement, with *Erica cinerea* (bell-heather), *Luzula multiflora* (many-headed woodrush) and *Vaccinium myrtillus* (bilberry) as selective species. The canopy is rather open, of oak or birch, and there is rarely an understorey or saplings present. The ground cover is usually *Pteridium aquilinum* (bracken), *Deschampsia flexuosa* (wavy hair-grass) or *Calluna vulgaris* (ling). The type covers a range of traditional descriptions but is mainly birch or oak woodland in poor, freely drained conditions. The soils are mainly brown podzolics or podzols.

29. *Anthoxanthum odoratum/Agrostis tenuis* (sweet vernal-grass/common bent-grass) type
A type of medium heterogeneity and an average species complement, with *Rumex acetosa* (sorrel), *Potentilla erecta* (common tormentil) and *Galium saxatile* (heath bedstraw) being the selective species. The canopy is usually of average density, of oak or birch, with saplings and an understorey often present. The ground cover is usually *Agrostis tenuis* (common bent-grass), *Pteridium aquilinum* (bracken) or *Holcus mollis* (creeping soft-grass). The type would probably be called upland oak or birch woodland. The soils are variable, but are mainly acid brown earth or brown podzolic in character.

SUMMARY OF **SITE TYPE 14**
ANTHOXANTHUM ODORATUM/QUERCUS (SWEET VERNAL-GRASS/OAK) TYPE

VEGETATION

Key species
Constant species: *Pteridium aquilinum* (bracken), *Anthoxanthum odoratum* (sweet vernal-grass), *Deschampsia flexuosa* (wavy hair-grass), *Galium saxatile* (heath bedstraw)

Plot dominants: *Pteridium aquilinum* (bracken), *Deschampsia flexuosa* (wavy hair-grass), *Agrostis tenuis* (common bent-grass), *Vaccinium myrtillus* (bilberry)

Selective species: *Anthoxanthum odoratum* (sweet vernal-grass), *Galium saxatile* (heath bedstraw), *Agrostis canina* (brown bent-grass), *Deschampsia flexuosa* (wavy hair-grass), *Vaccinium myrtillus* (bilberry), *Rhytidiadelphus squarrosus*

Blend of   Frequency 25, 26, 27, 29 (30)
plot types: Mean number 5.4 (med)

Mean number of species: 115 (med)
Total number of species: 218 (med)

Canopy and understorey species
Constant trees   *Constant saplings*
Birch (oak)   —

Constant shrubs   *Trees (basal area)*
—   Oak (pine, birch) (low density)

ENVIRONMENT

Geographical distribution	*Solid geology*	*Rainfall (cm)*
(WS, NW, Wa)	Silur/Ordov (J)	155 *(high)*

Altitude (m)	*Altitude (bot)*	*Altitude (top)*	*Slope (°)*	*Soil (pH)*	*LOI*
152 *(med)*	111 m *(high)*	268 m *(med)*	33.0 *(high)*	4.3 *(low)*	28.3 *(med)*

SITE TYPE 15

SUCCISA PRATENSIS/BETULA (DEVIL'S-BIT SCABIOUS/BIRCH) TYPE

A variable type most closely related to types 3 and 11, the former being much more lowland in character, although the soils are comparable, whereas the latter is usually associated with larger, more complex sites containing more extensive areas of woodland vegetation. In terms of overall species content this was the richest type in the series. Plot type 30 [*Succisa pratensis/Holcus mollis* (devil's-bit scabious/creeping soft-grass)] occurs most widely but types 26 [*Potentilla erecta/Holcus mollis* (common tormentil/creeping soft-grass)] and 27 [*Calluna vulgaris/Pteridium aquilinum* (ling/bracken)] are also locally common, with other types occurring in some areas. The vegetation also has a complement of some more northern species, eg *Trientalis europaea* (chickweed wintergreen). The soils are mainly gleyed brown earths, but gleys and acid brown earths are also present in appropriate topographic locations.

The main canopy species is birch, which often forms a pure canopy, but rowan is also scattered throughout, and other species such as alder and willow occur by the streams which pass through the sites. A variety of habitats and species are associated with these watercourses and many seepages and flushes are also present. The woods are often small and have steep banks dropping down to the streams mentioned above.

The woods are mainly surrounded by pasture but rough grazing, scrub and occasionally arable land are also present. The woods are not usually well fenced and are often grazed by domestic stock, mainly sheep. Roe deer are present in most sites. Regeneration is thus restricted but birch is quite common in protected places, and willow and rowan are present locally.

The woods usually occupy steep slopes dropping down to small streams or occasionally steep gorges or seepages lines on steep slopes, but, being small, occupy a limited altitudinal range. They have been haphazardly exploited, mainly for firewood, because the stands are usually rather too low-growing for useful timber trees. They do, however, often have an important function as protection for stock, and occasionally have extensive conifer plantations nearby.

Although the surveyed sites were in Scotland, this type is likely to be found widely in the uplands of western Britain, and there is some evidence that outliers may even be present in the lowlands.

Within the *Nature conservation review*, sites such as Glen Tarff (west Scotland), Coed Crafnant (north Wales) and Hellbeck and Swindale (north-west England) are likely to be included in this type.

On the Continent, such sites are not likely to be common but are mostly likely in western Norway, Brittany, the Massif Central and the foothills of the western Pyrenees.

In general terms, the sites in this type would probably be called small upland birch woods by enriched streamsides with a complex ground vegetation dominated by tall herbs. The main phytosociological association is probably Betulo-Quercetum Tx 1937 but other herb-rich types, eg Melico-Betuletum (KL pers. comm.), are probably also present.

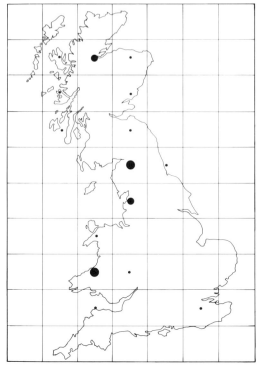

A wood in north-east
Scotland, with dense
woodland of *Alnus glutinosa*
(alder) and *Salix* spp. (willow)
by the riverside grading into
Betula spp. (birch) on the
steep upper slopes.

Type 15.   *Succisa pratensis/
Betula* (devil's-bit scabious/
birch)

30. *Succisa pratensis/Holcus mollis* (devil's-bit scabious/creeping soft-grass type
A very heterogeneous type with a high species complement, *Ranunculus repens* (creeping buttercup), *Veronica officinalis* (common speedwell) and *Geranium sylvaticum* (wood cranesbill) being selective species. The canopy is mostly open, usually of birch but also sometimes rowan. An understorey is often present, but saplings are in low density. The ground cover is usually of *Holcus mollis* (creeping soft-grass), *Pteridium aquilinum* (bracken) or *Agrostis tenuis* (common bent-grass). The type would probably be called herb-rich upland birch woodland. The soils are mainly gleys and are usually flushed being often beside streams.

27. *Calluna vulgaris/Pteridium aquilinum* (ling/bracken) type
A type of medium heterogeneity and an average species complement, with *Erica cinerea* (bell-heather), *Luzula multiflora* (many-headed woodrush) and *Vaccinium myrtillus* (bilberry) as selective species. The canopy is rather open, of oak or birch, and there is rarely an understorey or saplings present. The ground cover is usually *Pteridium aquilinum* (bracken), *Deschampsia flexuosa* (wavy hair-grass) or *Calluna vulgaris* (ling). The type covers a range of traditional descriptions but is mainly birch or oak woodland in poor, freely drained conditions. The soils are mainly brown podzolic or podzols.

26. *Potentilla erecta/Holcus mollis* (common tormentil/creeping soft-grass) type
A type of medium heterogeneity and an average species complement, with *Galium saxatile* (heath bedstraw), *Anthoxanthum odoratum* (sweet vernal-grass) and *Deschampsia flexuosa* (wavy hair-grass) as selective species. The canopy is usually of average density, consisting mainly of oak but with birch and rowan often present, as well as saplings and an understorey of hazel. The ground cover is usually of *Holcus mollis* (creeping soft-grass), *Pteridium aquilinum* (bracken) or *Agrostis tenuis* (common bent-grass). The type would usually be termed western acid sessile oak woodland but has some enrichment. The soils vary in character from acid brown earths to brown podzolic series.

22. *Blechnum spicant/Rubus fruticosus* (hard fern/bramble) type
A type with medium heterogeneity and an average species complement, *Luzula sylvatica* (greater woodrush), *Athyrium filix-femina* (lady-fern) and *Hedera helix* (ivy) being selective species. The canopy is of average density, mainly of oak but with birch, rowan, beech, ash and sycamore also present, and a few saplings and some shrubs. The ground cover is mainly *Rubus fruticosus* (bramble) and *Luzula sylvatica* (greater woodrush), but there is often a dense carpet of leaves. The type would usually be called mixed deciduous woodland on steep valley sides. The soils are mainly acid brown earths and often very rocky.

SUMMARY OF **SITE TYPE 15**

SUCCISA PRATENSIS/BETULA (DEVIL'S-BIT SCABIOUS/BIRCH) TYPE

VEGETATION

Key species
Constant species: *Betula* spp. (birch), *Pteridium aquilinum* (bracken), *Sorbus aucuparia* (rowan), *Anthoxanthum odoratum* (sweet vernal-grass)

Plot dominants: *Pteridium aquilinum* (bracken), *Holcus mollis* (creeping soft-grass), *Luzula sylvatica* (greater woodrush), *Agrostis tenuis* (common bent-grass)

Selective species: *Succisa pratensis* (devil's-bit scabious), *Luzula multiflora* (many-headed woodrush), *Lathyrus montanus* (bitter vetch), *Hylocomium splendens*, *Rhytidiadelphus triquetrus*, *Pseudoscleropodium purum*

Blend of   Frequency 30, 27, 26 (22, 28, 31)
plot types: Mean number 4.5 (low)

Mean number of species: 158 (high)
Total number of species: 251 (high)

Canopy and understorey species
Constant trees   *Constant saplings*
Oak   —

Constant shrubs   *Trees (basal area)*
(Hazel) (low density)   Oak (open canopy)

ENVIRONMENT

Geographical distribution	*Solid geology*	*Rainfall (cm)*
ES (WS)	Ign/Metam (I)	99 *(low)*

Altitude (m)	*Altitude (bot)*	*Altitude (top)*	*Slope (°)*	*Soil (pH)*	*LOI*
131 *(med)*	112 m *(high)*	389 m *(high)*	23.0 *(med)*	5.1 *(med)*	21.2 *(low)*